Consumption Motives in Luxury Marketing

An Analysis of two Agricultural Markets:

Equestrian Sports and Food

Consumption Motives in Luxury Marketing

An Analysis of two Agricultural Markets:

Equestrian Sports and Food

Dissertation

to obtain the doctoral degree

in the Faculty of Agricultural Sciences,

Georg-August-University of Goettingen, Germany

presented by

Laura Helena Hartmann

born in Leer (Ostfriesland)

Goettingen, May 2015

Bibliografische Information der Deutschen Nationalbibliothek

Die Deutsche Nationalbibliothek verzeichnet diese Publikation in der
Deutschen Nationalbibliografie; detaillierte bibliografische Daten sind im Internet
über http://dnb.d-nb.de abrufbar.
1. Aufl. - Göttingen: Cuvillier, 2015
   Zugl.: Göttingen, Univ., Diss., 2015

D7

1. Supervisor: Prof. Dr. Achim Spiller

2. Co-supervisor: Prof. Dr. Klaus-Peter Wiedmann

Date of dissertation: 21 May 2015

   ISBN 978-3-7369-9038-8
   eISBN 978-3-7369-8038-9

Contents

Short summary

This dissertation is devoted to attitudes of consumers toward luxury in two agricultural markets, horse sports and foods. Literature postulates a change of perceived luxury definitions and motives for luxury consumption. Accordingly, personally-oriented luxury consumption has gained significance while socially-oriented motives have been pushed into the background. Based on this, the following studies were aimed to reveal how far it has affected the consumer behavior in both agricultural markets. The research results are used to define the target groups for different kinds of luxury marketing and to give recommendations for the design of accordant marketing strategies. Thus, this work helps to fill a vacuum in marketing science: The value change in luxury consumption is a recent phenomenon through which new luxury markets have established and a need for new consumer studies in luxury marketing is created. Due to the timeliness of these developments, there is a lack of empirical target group specification in particular luxury markets that are focused on the differentiation between traditional and modern luxury consumption patterns. By investigating German and Chinese horse riders, the added aim of the dissertation was to reveal potential differences in attitudes toward luxury between cultures and to test the validity of modern (Western) consumption motives in a Confucian country. An excursion was used to screen applied marketing instruments or to reveal the success factors for marketing in an environment that is associated with luxury. Altogether, two studies analyze consumption motives in German and Chinese horse sports, three studies investigate the German market for luxury foods or upscale foods and two studies represent the excursion.

The analyses provide empirical evidence for the existence of an overall shift of motives for luxury consumption and luxury definitions away from prestige and conspicuousness toward self-realization, hedonism, intangible values, functionality, sustainability and authenticity. Moreover, the results imply that luxury consumption can be categorized in tangible luxury goods and *luxury experience*. Despite intersections by means of hedonism and self-realization, the studies revealed differences in the consumption motives for both categories of luxury.

Kurzzusammenfassung

Die vorliegende Dissertation untersucht die Einstellungen von Konsumenten zu Luxus in zwei agrarwirtschaftlichen Märkten – Pferdesport und Lebensmittel. In der Literatur wird das Postulat aufgestellt, dass im Luxuskonsum eine Entwicklung stattgefunden hat, nach der sich individuelle Luxusdefinitionen und Konsummotive in den letzten Jahrzehnten gewandelt haben. Demzufolge stehen persönlichkeitsorientierte Luxusmotive inzwischen im Vordergrund, während sozialorientierte Luxusmotive immer unwichtiger werden. Die folgenden Studien geben Aufschluss darüber, inwieweit dieser Wandel des Konsumentenverhaltens auf den Märkten für Pferdesport und Luxus-Lebensmittel empirisch nachgewiesen werden kann. Die Ergebnisse werden genutzt, um Zielgruppen auf Luxusmärkten zu definieren und Implikationen für das Marketing abzuleiten. Damit dient diese Arbeit zur Füllung eines Vakuums im Bereich der Erforschung neuer Luxusmotive: Der Wertewandel im Luxuskonsum ist ein aktuelles Phänomen, wodurch die stetige Entwicklung neuer Luxus-Nischenmärkte zu beobachten ist. Vor diesem Hintergrund verlangt die zielgruppenspezifische Ausrichtung von Marketingstrategien die Bereitstellung neuer, auf den Wandel ausgerichteten Erkenntnisse über das Konsumentenverhalten auf Luxusgütermärkten. Empirische Untersuchungen dazu wurden aber bislang wenig durchgeführt. Durch die Inklusion von Analysen im deutschen und chinesischen Reitsport liefert diese Dissertation zudem einen Überblick über Differenzen zwischen den Luxuseinstellungen in einer westlichen und einer konfuzianischen Kultur. Es wird eruiert, inwieweit die durch den Wandel hervorgebrachten Motive auch den konfuzianischen Luxuskonsum prägen. Eine Exkursion untersucht schließlich angewendete Marketinginstrumente im luxusassoziierten Umfeld des deutschen Reitsports.

Insgesamt beschäftigen sich zwei Studien mit den Konsummotiven im deutschen und chinesischen Reitsport, drei Studien analysieren den deutschen Markt für Luxus-Lebensmittel bzw. für hochpreisige Lebensmittel, und zwei Studien bilden die Exkursion. Die Analysen liefern im Allgemeinen empirische Beweise für den postulierten Wandel. Motive und Luxusanschauungen werden aus Konsumentensicht vielmehr über Selbstverwirklichung, Hedonismus, immaterielle Werte, Funktionalität, Nachhaltigkeit und Authentizität definiert als über Prestige und demonstrativem Konsum. Überdies implizieren die Ergebnisse, dass eine Kategorisierung von Luxus in tangible Luxusgüter und *Luxus-Erleben* sinnvoll ist. Dementsprechend ergeben sich neben einer Schnittmenge in Form von Hedonismus und Selbstverwirklichung Unterscheidungen in den ausschlaggebenden Konsummotiven.

Introduction

Today's luxury markets are flourishing. Market studies show that luxury material goods and luxury services are increasingly demanded internationally and researchers prognosticate a further growth of the respective markets (Bellaiche et al., 2012; D´arpizio & Levato, 2014). In 2012, the global luxury market volume was estimated at around 1.42 trillion Euros, and 14.9% of it was based on personal luxury-like apparel, accessories and perfumes and 46.8% on so-called *luxury experiences* like activities at auction houses, hotels and exclusive trips. A share of 38.3% belongs to other luxury items, where the category of food, wines and spirits, for example, contribute to 5.4% of the market (76 billion Euro-market volume) (Müller-Stewens, 2013). The average annual growth rate of the market for personal luxury goods, luxury cars and experiential luxury between 2010 and 2012 was approximately 13% (Bellaiche et al., 2014). D´arpizio and Levato (2014) reveal that at the end of 2013, the global luxury market numbered approximately 330 million consumers, which is more than threefold the amount of customers twenty years before.

These figures implicate the significance of luxury marketing strategies on today's international markets. The creation of effective, target-group specific instruments would enable producers of luxury goods and services to address broad consumer bases. In order not to fail, the generation of knowledge about the luxury consumers' behavior is a prerequisite (Ascheberg, Meurer, & Oesterling, 2012; Kisabaka, 2001; Mueller & Koch, 2012). A challenging step is to reveal the perceived definitions of luxury. Peoples' fascination for luxury is an epoch-spanning and cross-cultural phenomenon. It has been observed through all periods of human history and in all Western and Eastern cultures (Dubois, Laurent, Czellar, 2001; Lu, 2011). Among periods and cultures, luxury consumption patterns differ quite widely, since luxury is a latent, immaterial construct (Godey et al., 2013). Looking at the micro level of markets, the answer to the question "what does luxury actually mean" is even dependent on a particular consumer (Phau & Prendergast, 2000). Perceived definitions are formed by individual experiences, values and attitudes, which in turn are influenced by societal developments and value systems. This causes heterogeneous and dynamic luxury markets (Meurer & Manninger, 2012). The characteristics of luxury target groups are dependent on various variables such as time, region and the particular luxury product or service. Against this background, the development of marketing strategies that are tailored to the demands of customers on luxury markets requires intense market research (Ascheberg, Meurer, & Oesterling, 2012).

In congruence with the dynamic character of luxury markets, a shift in motives for Western luxury consumption can actually be observed. Traditional motives like prestige, social distinction and conspicuousness (Veblen, 1899) take a backseat to modern motives like indulgence, immaterialism, and self-realization. Consuming any luxury good or service or accumulating items of property does not only serve the externally oriented self-portrayal. Instead, consumers are increasingly satisfying their internal longing by means of luxury more frequently. The demand for hedonistic actions and individuality becomes a superficial motive for a delightful, luxurious lifestyle in the emancipated affluent society (e.g. Bellaiche et al., 2012; Yeoman, 2011; Yeoman & McMahon-Beattie, 2010). In this way, luxury consumers can match the modern paradigm of diagonal social integration. The objective here is to become an example within the social environment by means of living an auspicious and successful life. This concept stands for the societal admiration of individuality in lifestyles (Mohr, 2014).

In a face-to-face interview that we conducted with Gerd Kaefer, who founded the giant in delicatessens, restaurants and catering *Feinkost Kaefer* and can rely on many years of practical experiences with luxury markets and luxury target groups, confirms the results from research. He stated that "[...] *Luxury nowadays is an integrated overall concept. It is not only the food or the clothes [...], and under no circumstances it is showing-off and conspicuousness. It is more defined by a unique lifestyle of indulgence, elegance, individuality and success. Calling a lifestyle or a particular situation luxury means that every single component should be luxury, whereas the highest level of luxury can only be reached by means of the expression of personality, individuality and uniqueness [...].*"

By contrast to the diagonal social integration, the vertical social integration, where status symbols and demonstrations of wealth are used to signal membership to higher social classes, becomes a less attractive societal aim. Horizontal social integration is the third category of different possibilities for the social effectiveness of consumption styles. It stands for the establishment of subgroups within a society and thus causes the existing variety of lifestyles. The aim is to distinguish from others through a specificity that is common to the members of such a group. Non-hierarchical differentiation should be reached by the development of distinct (life) styles. This does not result in the stratification of a social structure; it leads instead to a horizontal stratification of tastes in a society (Mohr, 2014).

Some interconnected social developments are considered to cause this change in luxury consumption. On the one hand, a so-called "democratization of luxury consumption" can be observed in industrialized countries due to the increasing prosperity across social layers (Atwal & Williams, 2007; Vigneron & Johnson, 2004). On the other hand, individualism and the emancipation of women are associated with a shift from collective and masculine societal values toward the enhanced appreciation of individuality, self-actualizing and the detachment from standards (Abramson & Inglehart, 1995; Ahuvia, 2002). For example, the observed feminine, aesthetical imprint of some areas of consumption can be linked to these developments (Jantzen, Østergaard, & Vieira, 2006). Luxury represents an example for an area of consumption that is influenced by a female-dominated customer base. Women generally have a more positive attitude toward luxury brands than men[1] and perceive uniqueness and quality as important luxury dimensions (Stokburger-Sauer & Teichmann, 2013). Wiedmann, Hennigs, and Siebels (2009), as well as Vigneron and Johnson (2004), find evidence that both motives are parts of a multidimensional approach to define luxury from a consumer perspective; they are thus significant motives in modern luxury consumption. Yeoman (2011) states that male trophies and status symbols have been replaced by feminine motives of experience and indulgence in modern luxury consumption.

Figure 1 visualizes the findings on the value change in luxury consumption.

---

[1] Wang and Griskevicius (2014) find that luxury plays an important role for mate guarding. A woman's consumption of luxury goods signals to other women that her romantic partner is devoted to her and thus provides an instrument to detect her romantic relationship against threats by female rivals. Mate retention is historically more concerned by women than by men. For example, women incur higher reproductive costs when relationships break or their romantic partners do not provide (Geary, 2000; Hurtado & Hill, 1992).

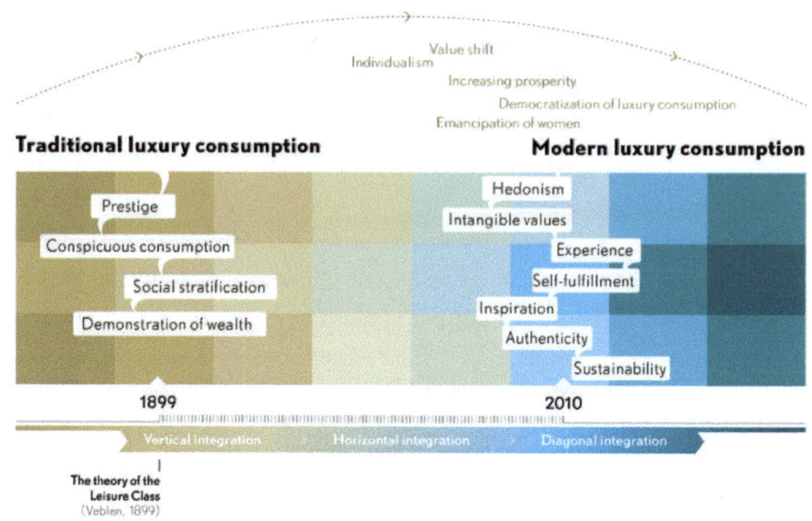

Figure 1. The change in motives for luxury consumption (own diagram)

Even though the need for studies on the change in motives for luxury consumption is a postulated manifold (Beverland, 2005; Vigneron & Johnson, 2004; Yeoman & McMahon-Beattie, 2006), respective empirical investigations are rare. Wiedmann, Hennigs and Siebels (2009) refer to general luxury brands and empirically develop a framework for the perceived dimensions of a luxury product value. It is suitable for the extension toward the postulated new luxury values and thus should be validated through studies on particular luxury markets or the appliance of different statistical methods.

In this dissertation, five different studies empirically investigate the consumption motives of two agricultural markets, horse sports and the market for foods. We conducted two of them on German and Chinese horse sports markets, and two of them focus on the market for luxury foods. The latter are introduced by a study that approaches the empirically hardly explored market for luxury food by empirically revealing expectations of consumers toward upscale food products. Hence, luxury food is here defined by means of its price level. *Price* is known to influence the perceived luxury level of a good (Dubois, Laurent, & Czellar, 2001; Wiedmann, Hennigs & Siebels, 2007, 2009), and in comparison to other luxury attributes the price level is objectively to ascertain. The results of this study are used as a first overview over the factors that influence consumer behavior in the market for luxury food. Furthermore, it inspires the research design of the following studies on the perceived dimensions of luxury food.

Both the horse sports markets and the market for luxury food provide research areas where traditional, as well as modern, motives for luxury consumption can be monitored. As far as I know, this opportunity has not been taken by other authors. The research questions of this dissertation have not been investigated before.

The choice of the market for horse sports and the market for foods was based particularly on arguments that I briefly outline here. German markets for equestrian products and services are significant in size in terms of stakeholders and turnovers. In 2013, 3.98 million people in Germany participated in horseback riding often or sometimes, and the overall turnover of the German equine industry was more than 5 billion Euros (AWA, 2013; D. R. Vereinigung, 2013). Ever since, horse sports have been associated with luxury.[2] High prices as well as the variety and high qualities of products and services in the market lead to the theory that equestrian riders generally have a high willingness to pay for their sports. Ikinger et al. (2013) find that the target group of equestrians is characterized by a high involvement in their sports, and the market studies conducted by Institut für Demoskopie Allensbach (2013) and IPSOS (2003) show that high income classes and a high level of education are predominant among the riders. Additionally, empirical studies in sports sciences reveal that the motives for doing horse sports correspond to modern motives for luxury consumption. Immaterial, internal values like enjoyment, experience, self-actualization, health and hedonism are perceived to be more important than classical motives like success and prestige. Even though, both modern and classical motives are significant (Gille, Hoischen-Taubner, & Spiller, 2011; Ikinger, Muench, Wiegand, & Spiller, 2013; IPSOS, 2003).

The Chinese market for horse sports has recently been set up at a high pace. Therefore, it has hardly been scientifically investigated yet. During my activities in Chinese horse sports and on a research journey to Beijing and Shanghai in the summer of the year 2013, I could observe that there is a growing interest in horse sports among affluent Chinese people. A high and steadily growing market turnover for stables, equestrian services and equipment for horse sports suggests that Chinese horse sports offer another example for a branch where strategies for *luxury and luxury experience* can be applied. However, China is known as one of the most important market places for luxury goods and services worldwide. Chinese consumers currently represent approximately 30% of the global luxury market (D´arpizio, 2014). Thus, I can

---

[2] See for the mentioning of horses in conjunction with luxury and prestige. Ammon (1828) and Buchner (1990).

assume that Chinese people have a general affinity toward luxury and can be addressed by luxury marketing strategies. The results of this dissertation for the consumer behavior of Chinese equestrians regarding luxury are based on interviews, a questionnaire and perennial observations in Chinese horse sports. The interviews and questionnaire were conducted during my research journey. In combination with a case study on the Chinese market for customized high-quality riding saddles, a marketing strategy can be designed that is inspired by one-to-one marketing (Peppers & Rogers, 1993, 1997; Peppers, Rogers, & Dorf, 1999).

The inclusion of one Western and one Confucian society in the studies of this dissertation enables me to explore cultural differences in luxury consumption and to compare recent trends. Germany and China provide two particular examples for internationally significant, actually growing, market places for luxury goods and services (Meurer, 2012; Zhan & He, 2012).

For the German food market, studies show the existence of a segmental change in patterns of consumption. This has shifted the consumers' motives toward similar consumption motives as those observed in Western luxury markets, in accordance with, e.g. Dubois, Czellar, and Laurent (2001), Meurer and Manninger (2012); Vigneron and Johnson (2004) and Wiedmann, Hennigs and Siebels (2007, 2009). Studies show that the food quality and health aspects, origin and production methods as well as indulgence or self-actualization and sustainability concerns become increasingly important among some consumer segments (Kirig & Ruetzler, 2007; Nestlé Deutschland AG, 2012; SGS, 2014). The development toward a growing awareness for food quality, health and sustainability has produced differentiated demands of consumers in the food market (Nitzko & Spiller, 2014). This trend has, for example, established the Lifestyle of Health and Sustainability (LOHAS). It denotes the aim to bring the individual lifestyle in line with sustainable, health- and indulgence-oriented consumption. This means that this target group buys expensive sustainable foods like organic or regional products rather than cheaper products without any sustainability, quality or health value. Meanwhile, every seventh consumer in the food market belongs to the group that subscribes to the ideas of LOHAS. The size of this target group has increased by almost 50% since 2007 (GFK, 2013).

At the same time, premium foods, which are associated with uniqueness, high quality and high prices, are becoming increasingly more important for the German food market. According to a recent study, 8.22 million German consumers have bought premium or deli-food items in the last 14 days (Institut für Demoskopie Allensbach, 2014). The consumer segments with higher income seem to have an especially high willingness to pay for food items. Instead of low prices, the purchase intention in the market for premium foods is based on product competencies and special ways of cultivation or production (Nitzko & Spiller, 2014). In summary, these findings implicate that a food market exists that is strongly associated with the dimensions of luxury. Kirig & Ruetzler (2007) state that foods are the luxury market of the future.

In addition to the investigation of perceived definitions of luxury and motives for its consumption in two agricultural markets, the dissertation includes an excursion, represented by two additional studies on the horse sport market. It addresses marketing problems in a luxury-associated branch. One study provides enhancements of the ranking system for jumping horses by means of an experiment, and the other study empirically reveals success factors for equestrian tourism.

The first part of this dissertation contains the studies on horse sports; the second part deals with luxury foods and the excursions are located in the third part. The concluding section finally summarizes the findings of the seven research articles and discusses them with regard to implications for marketing and research. Furthermore, it evaluates the limitations of the dissertation from which further need for research is derived.

References

Abramson, P., & Inglehart, R. F. (1995). *Value change in global perspective*. University of Michigan Press.

Ahuvia, A. C. (2002). Individualism/collectivism and cultures of happiness: A theoretical conjecture on the relationship between consumption, culture and subjective well-being at the national level. *Journal of Happiness Studies, 3*(1), 23-36.

Ammon, G. G. (1828). *Ueber die Eigenschaften des Soldaten-Pferdes und die Mittel, die Zucht desselben zu befördern*. Mittler.

Ascheberg, C., Meurer, J., & Oesterling, A. (2012). The Luxury Universe–Angebots-und Kundensegmentierung globaler Luxusmärkte als Basis für erfolgreiche Positionierungsstrategien. In C. Burmann, V. König, & J. Meurer (Hrsg.), *Identitätsbasierte Luxusmarkenführung* (S. 85-101). Springer Fachmedien Wiesbaden.

Atwal, G., & Williams, A. (2007). Experiencing luxury. *ADMAP, 481*, 30.

Bellaiche, J., Kluz, M., Mei-Pochtler, A., & Wiederin, E. (2012). *Luxe Redux: raising the bar for selling of luxuries*. Boston Consulting Group (Ed.), Boston.

Beverland, M. B. (2005). Crafting brand authenticity: the case of luxury wines. *Journal of Management Studies, 42*(5), 1003-1029.

Buchner, J. (1990). Von Pferden, Hühnern und Läusen. In Behnken, I. (Hrsg.) (S. 90-97). *Stadtgesellschaft und Kindheit im Prozeß der Zivilisation.* VS Verlag für Sozialwissenschaften.

D. R. Vereinigung e.V. (2013). *Jahresbericht 2013.* FN-Verlag.

D´arpizio, C. & Levato, F. (2014). *Lens on worldwide luxury consumer: Relevant segments, behaviors and consumption patterns, nationalities and generations compared.* Bain and Company (Ed.), Boston.

D´arpizio, C. (2014). *Global Luxury Goods Worldwide Market Study Spring 2014.* Bain and Company (Ed.), Boston.

Dubois, B., Laurent, G., & Czellar, S. (2001). Consumer rapport to luxury: Analyzing complex and ambivalent attitudes. *Le Cahiers de Recherche, 736,* HEC Paris.

Geary, D. C. (2000). Evolution and proximate expression of human paternal investment. *Psychological bulletin, 126*(1), 55.

GFK Consumer Panels und Bundesvereinigung der Deutschen Ernährungsindustrie (2013) (Hrsg.). Consumers´ Choice ´13: Bewusster Genuss-Nachhaltige Gewinne für Ernährungsindustrie und Konsumenten. 5. Ausgabe.

Gille, C., Hoischen-Taubner, S., & Spiller, A. (2011). Neue Reitsportmotive jenseits des klassischen Turniersports. *Sportwissenschaft, 41*(1), 34-43.

Godey, B., Pederzoli, D., Aiello, G., Donvito, R., Wiedmann, K. P., & Hennigs, N. (2013). A cross-cultural exploratory content analysis of the perception of luxury from six countries. *Journal of Product & Brand Management, 22*(3), 229-237.

Honkanen, P., Verplanken, B., & Olsen, S. O. (2006). Ethical values and motives driving organic food choice. *Journal of Consumer Behaviour, 5*(5), 420-430.

Hurtado, A. M., & Hill, K. R. (1992). Paternal effect on offspring survivorship among Ache and Hiwi hunter-gatherers: Implications for modeling pair-bond stability. In B. S. Hewlett (Ed.), *Father-child relations: Cultural and biosocial contexts* (p. 31-55). Aldine de Gruyter, New York.

Ikinger, C., Muench, C., Wiegand, K., & Spiller, A. (2013). Reiterleben, Reiterwelten: Zielgruppen zwischen Reitweisen, Motiven und der Liebe zum Pferd. Georg-August-Universität Göttingen, HorseFuturePanel UG, Dietz und Consorten (Hrsg.). URL: http://www.uni-goettingen.de/de/document/download/

1988e74b5e6a7bf92bf38381a71a47f0.pdf/2013-04%20reitsportstudie_screen.pdf. Zugriff am 05.Mai 2013.

Institut für Demoskopie Allensbach (2013). *Allensbacher Markt- und Werbeträgeranalyse 2013*. Allensbach: Institut für Demoskopie Allensbach.

Institut für Demoskopie Allensbach (2014). *Allensbacher Markt- und Werbeträgeranalyse 2014*. Allensbach: Institut für Demoskopie Allensbach.

IPSOS (2003). *Faszination Zukunft. Neue Perspektiven im Pferdesport. Die FN Marktanalyse kompakt und kommentiert*. Warendorf: FN Verlag.

Jantzen, C., Østergaard, P., & Vieira, C. M. S. (2006). Becoming a 'woman to the backbone': Lingerie consumption and the experience of feminine identity. *Journal of Consumer Culture, 6*(2), 177-202.

Kirig, A., & Ruetzler, M. H. (2007). Food-Styles. *Die wichtigsten Thesen, Trends und Typologien für die Genuss-Märkte*. Zukunftsinstitut-Studie. Kelkheim.

Kisabaka, L. (2001). *Marketing für Luxusprodukte* (Vol. 32). Dissertation, Fördergesellschaft Produkt-Marketing, Cologne.

Lu, P. X. (2011). *Elite China: luxury consumer behavior in China*. John Wiley & Sons.

Meurer, J. (2012). Ebony or Ivory–wie glänzend ist die Zukunft des Luxus in Deutschland? Kritische Reflexionen zum Luxusmarkenmanagement. In C. Burmann, V. König, & J. Meurer (Hrsg.), *Identitätsbasierte Luxusmarkenführung* (S. 321-336). Springer Fachmedien Wiesbaden.

Meurer, J., & Manninger, K. (2012). Quo vadis globale Luxusmarkenführung–Status, Trends und Top-Themen für die CMO-Agenda. In C. Burmann, V. König, & J. Meurer (Hrsg.), *Identitätsbasierte Luxusmarkenführung* (S. 321-336). Springer Fachmedien Wiesbaden.

Mohr, E. (2014). *Ökonomie mit Geschmack: die postmoderne Macht des Konsums*. Murmann Verlag DE.

Müller, F., & Koch, K. D. (2012). Erfolgreiches Luxusmarketing–Eine provokative Diskriminierung. *Marketing Review St. Gallen, 29*(1), 11-16.

Müller-Stewens, G. (2013). *Das Geschäft mit Luxusgütern*. Universität St. Gallen.

Nestlé Deutschland AG (2012) (Hrsg.). *Nestlé Studie 2012, Das is(s)t Qualität*, Auszüge aus der Nestlé Studie 2012.

Nitzko, S., & Spiller, A. (2014). Zielgruppenansätze in der Lebensmittelvermarktung. In M. Halfmann (Hrsg.), *Zielgruppen im Konsumentenmarketing* (S. 315-332). Springer Fachmedien Wiesbaden.

Peppers, D., & Rogers, M. (1993). *The One to One Future: Building Relationships One Customer at a Time*, New York: Currency/Doubleday.

Peppers, D., & Rogers, M. (1997). *The one to one future*. New York: Doubleday.

Peppers, D., Rogers, M., & Dorf, B. (1999). Is your company ready for One-to-one marketing. *Harvard Business Review*, 77(1), 151-160.

Phau, I., & Prendergast, G. (2000). Consuming luxury brands: the relevance of the 'rarity principle'. *The Journal of Brand Management*, 8(2), 122-138.

SGS (2014). Vertrauen und Skepsis: Was leitet die Deutschen beim Lebensmitteleinkauf? SGS-Verbraucherstudie 2014: Ergebnisse einer bevölkerungsrepräsentativen Befragung. Hamburg: SGS Germany GmbH.

Stokburger-Sauer, N. E., & Teichmann, K. (2013). Is luxury just a female thing? The role of gender in luxury brand consumption. *Journal of Business Research*, 66(7), 889-896.

Veblen, T. (1899). *The theory of the leisure class*. Oxford University Press.

Vickers, J. S., & Renand, F. (2003). The marketing of luxury goods: an exploratory study–three conceptual dimensions. *The Marketing Review*, 3(4), 459-478.

Vigneron, F., & Johnson, L. W. (2004). Measuring perceptions of brand luxury. *The Journal of Brand Management*, 11(6), 484-506.

Wang, Y., & Griskevicius, V. (2014). Conspicuous Consumption, Relationships, and Rivals: Women's Luxury Products as Signals to Other Women. *Journal of Consumer Research*, 40(5), 834-854.

Wiedmann, K. P., Hennigs, N., & Siebels, A. (2007). Measuring consumers' luxury value perception: a cross-cultural framework. *Academy of Marketing Science Review*, 7(7), 333-361.

Wiedmann, K. P., Hennigs, N., & Siebels, A. (2009). Value-based segmentation of luxury consumption behavior. *Psychology & Marketing*, 26(7), 625-651.

Yeoman, I. & McMahon-Beattie, U. (2010) The changing meaning of luxury. In I. Yeoman, & U. McMahon-Beattie (Eds.), *Revenue Management: A Practical Pricing Perspective*, (chapter 6, pp. 62–85). Basingstoke, UK: Palgrave MacMillan.

Yeoman, I. (2011). The changing behaviours of luxury consumption. *Journal of Revenue & Pricing Management*, 10(1), 47-50.

Yeoman, I., & McMahon-Beattie, U. (2006). Luxury markets and premium pricing. *Journal of Revenue and Pricing Management*, 4(4), 319-328.

Zhan, L., & He, Y. (2012). Understanding luxury consumption in China: consumer perceptions of best-known brands. *Journal of Business Research*, 65(10), 1452-1460.

Chapter I: Personally-Oriented and Socially-Oriented Luxury Motives in Horse Sports

I. 1   **Luxusaffinität deutscher Reitsportler – Implikationen für das Marketing im Reitsportsegment**

Autoren: **Laura Hartmann, Achim Spiller**

Georg-August-Universität Göttingen

Dieser Beitrag wurde in ähnlicher Form als Diskussionspapier der Georg-August-Universität Göttingen veröffentlicht (Diskussionspapier 1501, *Diskussionspapiere der Georg-August-Universität Göttingen, Department für Agrarökonomie und Rurale Entwicklung*, ISSN 1865-2697).

Zusammenfassung

Der Reitsport wird seit jeher mit Luxus assoziiert. Das Durchschnittseinkommen und die Zahlungsbereitschaften von Reitsportlern sind hoch, zugleich wird ihnen eine Affinität zu Luxusgütern unterstellt. Dies erklärt das Interesse von Luxusgüterherstellern an einer Positionierung ihrer Marketingmaßnahmen im Reitsport. Motiviert durch die vielfache Verknüpfung beider Branchen wurde eruiert, inwieweit deutsche Reiter einem materiellen Luxusverständnis zugewandt sind. Diese Frage ist für Positionierungsstrategien im Luxusmarketing grundlegend, bisher aber kaum empirisch untersucht worden. Auf Grundlage einer Faktoren- und Clusteranalyse können vier Reitersegmente identifiziert werden, die sich in ihrer Haltung zu Luxus und ihrem sozio-demografischen Profil voneinander unterscheiden. Der Konsum von Luxus und die Ausübung von Reitsport werden eher motiviert durch das Bedürfnis nach Selbstverwirklichung und hedonistischem Erleben. Damit stellt die Gruppe der Reitsportler ein Anwendungsbeispiel für den in der Literatur zwar postulierten, bisher aber wenig empirisch untersuchten Wandel von außengerichteten, sozialen zu innengerichteten, persönlichkeitsorientierten Motiven im Luxuskonsum dar. Es können Implikationen für das Luxusgütermarketing abgeleitet werden.

Summary

Equestrian sports have always been associated with luxury. Equestrian athletes' average income and their willingness to pay are high. Simultaneously, they are assumed to have an affinity to luxury goods. This explains the interest of luxury goods manufacturers in placing their marketing activities in equestrian sports. Motivated by several links between both fields, this study found out how far German equestrian athletes have been turned to an understanding of material luxury. This is a fundamental question for positioning strategies in luxury marketing; however, it has hardly been studied empirically so far. Based on a factor and cluster analysis, four segments can be identified, differing in attitude as to luxury and socio-demographic profiles. Consuming luxury and exercising equestrian sports is rather motivated by the need for self-realization and hedonism, which means the group of riders are an application example of the change from external, social motives toward internal personality-related motives in luxury consumption. The results of the study on luxury marketing are assisting manufacturers and service providers in how to better serve (or service) the target group.

Schlüsselbegriffe

Luxusgütermarketing, Reitsportler, Luxusaffinität, Luxusmotive

Keywords

Luxury Marketing, Horse Riders, Affinity to Luxury, Luxury Motives

Executive summary

Der internationale Markt für Luxusgüter wächst stetig und verzeichnet seit einigen Jahren eine ungewöhnlich große Nachfrage. Auch für Deutschland wird dieser Trend bestätigt. Aufgrund ihrer Einkommens- und Bildungsstärke sowie hoher Zahlungsbereitschaften werden Reitsportler als geeignete Zielgruppe für Luxusgütermarketing identifiziert. Beispiele für die Verknüpfung der Luxusbranche mit dem Reitsport im Marketing liefern die Manufaktur für Mode, Lederwaren und Reitsportartikel *Hermes*, der Hersteller von (Reit-)Handschuhen *Roeckl*, der Uhrenhersteller *Rolex*, die Automobilmarke *Mercedes Benz* und die Modekonzerne *Gucci* und *Escada*. Praktiziert werden insbesondere Sponsoring-Aktivitäten auf Reitsportveranstaltungen, Werbeverträge mit prominenten Reitsportlern, die Verknüpfung von Produktlinien sowie die Verwendung von typischen Reitsportbildern bei der Gestaltung der Kommunikation.

Seit einiger Zeit wird ein Wandel in den Motiven des westlichen Luxuskonsums diskutiert. Traditionelle Konsummuster wie Prestige und soziale Distinktion treten vor modernen Motiven wie Hedonismus, Immaterialismus und Selbstverwirklichung in den Hintergrund. Der Konsum von teuren Marken und Dienstleistungen bzw. die Anhäufung wertvoller Vermögensgegenstände dient nicht mehr nur der außengerichteten Selbstdarstellung. Stattdessen – so die Theorie – befriedigen Konsumenten mithilfe von Luxus zunehmend innengerichtete Sehnsüchte. Das Bedürfnis nach hedonistischen Handlungen und Individualität wird in den emanzipierten Wohlstandsgesellschaften zum zentralen Motiv für eine genussvolle, luxuriöse Lebensart. Auch für den Reitsport werden immaterielle, hedonistische Motive als ausschlaggebend identifiziert.

Die vorliegende Studie untersucht, inwieweit deutsche Reitsportler einem materiellen Luxusverständnis zugewandt sind. Dabei zeigt sich, ob das Segment der Reitsportler als Zielgruppe für klassisches, auf das traditionelle Luxusverständnis ausgerichtete Luxusgütermarketing identifiziert werden kann oder ob es einen Anwendungsfall für den Wertewandel im Luxuskonsum darstellt. Diese Frage ist für Positionierungsstrategien im Luxusmarketing grundlegend, bisher aber kaum empirisch untersucht worden.

Anhand einer Befragung von 646 deutschen Reitsportlern wurden zwei Skalen zur Messung der Affinität zu Luxusprodukten und -aktivitäten angewendet und darüber hinaus soziodemografische Daten sowie Informationen über die Art und das Ausmaß der reitsportlichen Aktivitäten der Probanden erhoben. Es konnten vier Kundensegmente hinsichtlich des Kriteriums der Affinität zu einem materiellen Luxusverständnis gefunden werden. Dabei stellte sich heraus, dass in diesem Sinne nur etwa 11% der Probanden als entsprechend luxusaffin eingestuft werden können. Das entsprechende Segment ist einkommensstark und weist eine Vorliebe für Luxusprodukte bzw. teure Marken und exklusive Hobbys auf. Ein weiteres Segment bildet zwar ebenfalls einen positiven Faktor „Affinität zu Luxusprodukten", jedoch wird in diesem der Reitsport an sich mit Luxus assoziiert. Der Luxuskonsum konzentriert sich hier auf teure Pferde und optimale Reitsportbedingungen. Diese Reiter investieren gerne in die Qualität ihres Sports. Ihr Einkommen ist ebenfalls hoch, wenn auch niedriger als im ersten Segment. Ein Drittel der Befragten wird durch eine ambivalente Einstellung charakterisiert und ein weiteres Drittel hat kaum Interesse an teuren Konsumgütern oder eine ablehnende Haltung zu Luxus.

Allgemein kann festgestellt werden, dass die befragten Reiter ihren Sport regelmäßig und zu einem Großteil mit Wettkampfambitionen betreiben. 86% der Probanden besitzen eigene, ein oder mehrere Pferde. Der Reitsport nimmt einen ausgeprägten Stellenwert in ihrem Leben ein, zudem sind ihre Zahlungsbereitschaften hinsichtlich des Reitens und der Pferdehaltung sowie der betriebene Zeitaufwand hoch. Immaterielle Werte wie Gesundheit, Familienleben, Zeit für sich und die Pferde werden geschätzt und als „wahrer" Luxus empfunden. Das Reiten ist mehr als Lifestyle-Konzept aufzufassen denn als bloße sportliche Betätigung.

Die Analysen zeigen, dass sich im Reitsport praktiziertes Luxusmarketing nicht auf außengerichtete – traditionelle – Luxus-Konsummotive wie Prestige und Wohlstandsdemonstration konzentrieren sollte. Ein rein materielles und elitäres Luxusverständnis kann nur für ein relativ kleines Segment bestätigt werden. Stattdessen zeigte sich die Relevanz von innengerichteten Motiven, immateriellem Reichtum und ein spezifischer Lebensstil, charakterisiert durch Naturverbundenheit, Tierliebe, Verbindlichkeit und Genuss. Es wird deutlich, dass die Werthaltungen von Reitern den Wandel im allgemeinen Luxusverständnis widerspiegeln. Demzufolge bietet dieser Sport ein Szenario, aus welchem wichtige Erkenntnisse für das praktische

Luxusgütermarketing sowie für die Luxusforschung, insbesondere bezogen auf den Bereich des sogenannten *luxury experience*, gewonnen werden können.

Executive summary

The international market for luxury goods is continuously increasing. Even for Germany this trend can be confirmed. Due to their high income and education, as well as degree of willingness to pay, the equestrian athletes are identified as a target group for luxury goods marketing. Some examples of linking the luxury field with equestrian sports in marketing are found by the manufactory for fashion, leather goods and horse-riding articles, such as *Hermes*, the (riding) gloves producer *Roeckl*, the watchmaker Rolex, the car brand Mercedes-Benz and the fashion companies *Gucci*, *Chanel* and *Escada*. Marketing means, in particular, some sponsoring activities for equestrian events, advertising contracts with prominent equestrian athletes, the combination of product lines and the use of horses and other typical equestrian motives.

For some time a change in the motives of Western luxury consumption has been discussed. Traditional consumption patterns like prestige and social distinction take a backseat to modern motives like hedonism, immaterialism and self-realization. Consuming expensive brands and services or accumulating items of property does not only serve the externally-oriented self-portrayal. Instead, according to theory, consumers increasingly satisfy their internal longing by means of luxury. The demand for hedonistic actions and individuality becomes a superficial motive for a delightful, luxurious lifestyle in the emancipated affluent society. Even in equestrian sports immaterial, hedonistic motives are identified.

The present study investigates how far German equestrians' attitudes toward material understandings of luxury go. It answers the question if equestrians can be identified as a target group for the classical luxury goods marketing being focused on the traditional understanding of luxury or, if they present an application case for the change in luxury consumption motives. This is a fundamental question for positioning strategies in luxury marketing; however, it has hardly been studied empirically so far.

In an interview of 646 German equestrians, two scales for measuring the affinity to luxury—products and activities—have been applied and beyond, socio-demographic data as well as information on the nature and scope of the subjects' riding activities have been collected. Four customer segments concerning the affinity criterion to a material luxury understanding could be found. It came out that only ca. 11 % of the subjects can be seen as accordingly luxury affine. The corresponding segment has a high income and shows a preference for luxury products and expensive brands and exclusive hobbies, respectively. Another segment also forms a positive factor "Affinity to Luxury Products"; however, here equestrian sports themselves are associated with luxury. Consuming luxury concentrates on expensive horses and optimal equestrian sports conditions. These riders like investing in the quality of their sport. Their income is also high average, but lower than in the first segment. One-third of the sample is characterized by an ambivalent attitude and another third is scarcely interested in expensive luxury goods or has a reluctant attitude. Generally, it was found that the riders do their sport ambitiously. Ca. 86 % own one or more horses and equestrian sport is important in life. An extraordinarily high willingness to pay concerning riding as well as the expenditure of time is shown. Immaterial values like health, family life and time are appreciated and seen as "true" luxury. Riding is seen more as a lifestyle than a pure hobby. Analyses show that luxury marketing practiced in equestrian sports should not focus on external (traditional) luxury consumption motives. A purely material and elitist definition of luxury can only be confirmed for a relatively small consumer segment. Instead, the relevance of internally-related motives, immaterial richness, and a specific lifestyle being characterized by the attachment to nature, love for the animal, commitment and pleasure is evidenced. The riders' values thus reflect the change in the general understanding of luxury. Consequently, this sport offers a scenario providing important evidence for practical luxury-goods marketing as well as for luxury research, especially in the field of the so-called *luxury experience*.

Einleitung

Der internationale Markt für Luxusgüter verzeichnet seit einigen Jahren ein ungewöhnlich großes Nachfragewachstum (D´arpizio, 2014). Auch für Deutschland wird der Trend zu einer größeren Beliebtheit von Luxusartikeln bestätigt und eine Verstärkung dessen innerhalb der nächsten Jahre prognostiziert (Kewes, 2012; Meurer, 2012). Gleichzeitig wird eine Veränderung in den Motiven für Luxuskonsum in westlichen Ländern postuliert. Traditionelle Muster, wie der von Veblen (1899) thematisierte demonstrative Konsum, treten gegenüber Motiven wie Hedonismus, Immaterialismus und Selbstverwirklichung in den Hintergrund. Der Konsum von teuren Marken und Dienstleistungen bzw. die Anhäufung wertvoller Vermögensgegenstände dient nicht mehr nur der außengerichteten Selbstdarstellung. Stattdessen befriedigen Konsumenten mithilfe von Luxus zunehmend innengerichtete Sehnsüchte nach hedonistischen Handlungen, nach Erlebnis und Individualität (Ascheberg, Meurer, & Österling 2012; Stegemann, 2006; Vigneron & Johnson 2004; Yeoman, 2011). Dadurch, dass in den reichen Industrienationen materieller Luxus weitgehend schichtenübergreifend verfügbar ist, gewinnt die Befriedigung intrinsischer Bedürfnisse in Abgrenzung zu den sozialen Bedürfnissen möglicherweise an Bedeutung (Yeoman & McMahon-Beattie, 2006).

Kisabaka (2001) spricht in diesem Zusammenhang von einem Wertewandel, durch welchen selbstbezogene und stimulierende Werte in den Vordergrund rücken und zu Entscheidungskriterien auf dem Luxusgütermarkt werden. Sie schaffen eine kognitive Grundlage für das Bedürfnis nach Luxus und verdrängen dabei ökonomische und soziologische Kriterien der Kaufentscheidung als Erklärungsparameter für das Verhalten von Luxuskonsumenten. Luxusgüter werden weniger anhand ihres physischen Wertes beurteilt als vielmehr auf der Grundlage des mit ihnen assoziierten Zugewinns an emotionaler Befriedigung (Mostovicz, 2010). Dies stellt das Luxusmarketing vor Herausforderungen. In der unternehmerischen Praxis müssen neue Wege für die Kommunikation solcher Werte gefunden werden. Vor diesem Hintergrund sollten Konsumentenprofile erforscht und alte Marketing-Praktiken kritisch überprüft werden. Empirische Untersuchungen zum Luxusverständnis spezifischer Zielgruppen – hier von Reitern – wurden bis dato kaum durchgeführt. Weiterhin lieferte die Empirie bisher keine Antwort auf die Frage, ob und wie sich Luxusmotive in Bezug auf verschiedene Lebensbereiche – hier der Sport – unterscheiden. Fasst man individuelle Lebensstile als eine systematische Kombination komplementärer Konsumbereiche auf (Mohr, 2014), liegt die Annahme nahe, dass innerhalb dieser Bereiche unterschiedliche Motive wirken.

Es konnte mehrfach gezeigt werden, dass Sport stark von gesellschaftlichen Werten und Entwicklungen beeinflusst wird (Darlison, 2000; Norden & Polzer, 1995). Des Weiteren finden wir im Sport einen Bereich des menschlichen Zusammenlebens, in dem insbesondere immaterielle Werte gelebt werden. Mit Gleichgesinnten körperlich aktiv zu sein steht in engem Zusammenhang mit Gesundheit, Wohlbefinden und positiver sozialer Interaktion (Sudeck & Schmidt, 2012).

Der Reitsport im Speziellen bietet eine Projektionsfläche für die Betrachtung von Zielgruppen im Luxusmarketing. Eine Verknüpfung von Luxus und Reiten findet sich in der aktuellen Werbelandschaft sehr häufig, so z.B. bei Sponsoring-Aktivitäten von Luxusherstellern auf Reitsportveranstaltungen, bei Werbeverträgen mit prominenten Reitsportlern, bei der Zusammenführung von Produktlinien sowie bei der Verwendung von Pferde-Motiven und anderen typischen Motiven aus dem Reitsport bei der Gestaltung des Marketings. Bekannte Beispiele hierfür bieten die Manufaktur für Mode, Lederwaren und Reitsportartikel *Hermes*, der Hersteller von Handschuhen *Roeckl*, der Uhrenhersteller *Rolex*, die Automobilmarke *Mercedes Benz* und die drei Modekonzerne *Gucci, Chanel* und *Escada* (vgl. für den Fall *Rolex* Adjouri & Stastny, 2006).

Aus historischer Sicht werden das Pferd und der Reitsport schon seit jeher mit dem durch Prestige und soziale Stratifikation motivierten traditionellen Luxuskonsum in Verbindung gebracht.[3] Weiterhin führen sozio-ökonomische Eigenschaften von Reitsportlern wie ein erhöhtes verfügbares Einkommen und Bildungsnähe (Institut für Demoskopie Allensbach, 2013; IPSOS, 2003) dazu, dass sie als besonders luxusmarkenaffin eingestuft und als Zielgruppe klassischen Luxusgütermarketings wahrgenommen werden. Allerdings ist die Annahme der Affinität von Reitern zu traditionellem materiellen Luxus bisher nicht empirisch überprüft worden.

Aus aktuellen Studien geht hervor, dass Freude, Selbstverwirklichung, Gesundheit und Hedonismus zu den Hauptmotivatoren für die Ausübung von Reitsport gehören. Innengerichtete und immaterielle Motive werden bei Befragungen von Reitsportlern häufiger genannt als soziale und außengerichtete Motive wie Erfolg, Leistung und Anerkennung (Gille, Hoischen-Taubner, & Spiller, 2011; Häggblom, Rantamäki-Lahtinen, & Vihinen, 2012; D. R. Vereinigung, 2001). Auch die Kaufentscheidungen auf den Reitsportmärkten sind überwiegend an

---

[3] siehe für die Erwähnung von Pferden in Verbindung mit Luxus und Prestige z. B. Ammon (1828) und Buchner (1990)

hedonistischen und emotionalen Kriterien orientiert. Insbesondere die Preisbildung bei Pferden wird zu großen Teilen beeinflusst von Emotionalität und nicht- monetären Komponenten des Kaufgegenstands. Rationale Erwägungen finden dabei nur in reduziertem Maße Eingang in die Entscheidungsprozesse bei Nachfragern und Anbietern (Gamrat & Sauer, 2000). Daraus geht hervor, dass die Motivlage im Reitsport prima facie eng mit der Bewegungsrichtung des Wertewandels im Luxuskonsum korrespondiert. Auf beiden Märkten beobachten wir Konsummotive, die die traditionelle Annahme von ausschließlich prestige- und besitzorientierten Zielgruppen nicht länger stützen.

In der vorliegenden Studie wird deshalb die Affinität von deutschen Reitsportlern zu materiellem Luxus geprüft und auf Grundlage dessen eine Segmentierung in verschiedene Konsumentengruppen vorgenommen. Damit kann auch die Frage beantwortet werden, inwieweit sich die modernen, innengerichteten Reitsportmotive auf das Konsumverhalten dieser Gruppe im Luxusbereich auswirken. Weiterhin wird innerhalb der Segmente eine Charakterisierung des Luxusverständnisses vorgenommen. Für die Marketing-Praxis in Luxus- und Reitsportmärkten liefern die Ergebnisse ein Fundament für das Überdenken gewohnter Praktiken. Es können Handlungsempfehlungen für die Anpassung des Marketings an den modernen Luxusbegriff abgeleitet werden.

In Hinblick auf die deutsche Marketingforschung liefert die Studie erstmals empirische Ergebnisse zur Untersuchung des Luxuskonsums innerhalb einer Zielgruppe, die sich über die Ausübung eines gemeinsamen Sports definiert. Aufgrund der generellen Eigenschaft von Sport, gesellschaftliche Werte(-verschiebungen) abzubilden[4], sowie der Verknüpfung des Reitsports mit Luxusmärkten im Speziellen, findet sich hier ein Szenario, anhand dessen das Luxusverständnis segmentbezogen identifiziert werden kann. Weiterhin leistet die Studie einen Beitrag zur Entwicklung von Instrumenten zur Messung von Luxusaffinität innerhalb spezifischer Zielbranchen. In Anknüpfung an Dubois und Laurent (1994, 1995) und Dubois, Czellar und Laurent (2005) werden zwei zuvor entwickelte Skalen auf das Untersuchungsumfeld angepasst und kombiniert angewendet.

---

[4] siehe für die Verknüpfung von Sport mit gesellschaftlichen Werten und Entwicklungen Krüger (1988)

Konsummotive im Luxussegment

Für den Bereich des Marketings im Reitsport ist bislang generell nur wenig Literatur verfügbar. Auch im Bereich des Luxusmarketings bezogen auf Deutschland wurde bis dato wenig empirisch geforscht, woraus ein hoher Bedarf an Erkenntnisgewinn in diesem dynamischen Feld resultiert (Meurer, 2012). Die Forschungsbemühungen an den Schnittstellen beider Bereiche nehmen unseres Wissens einen völlig neuen Untersuchungsgegenstand in den Blick.

Luxus im Allgemeinen wird in der Literatur als vielschichtiges Konstrukt begriffen (Wiedmann, Hennigs, & Siebels, 2007, 2009). Das ursprüngliche lateinische Wort *luxus* steht für Extravaganz, Genuss, Reichtum und Üppigkeit (Glare, 1992), zugleich definiert die Marketingforschung den „Traumwert", das Verlangen nach dem Konsum eines bestimmten Luxusgutes, als maßgebendes Charakteristikum (Albrecht et al., 2013; Dubois & Paternault, 1995). Dennoch ist die Beschreibung der Güter, welche als Luxus bezeichnet werden, höchst subjektiv. Zunächst kann zwischen materiellen und immateriellen Luxus-Werten differenziert werden (Pflanz, 2004). Dabei erfolgt die Beschreibung des Konstrukts Luxus anhand von Konsummotiven. Diese unterliegen nicht nur einer kulturspezifischen Prägung, sondern sind darüber hinaus eng verknüpft mit dem zeitlichen Wandel von Wertesystemen (Chevalier & Lu, 2010; Wong & Ahuvia, 1998).

In Bezug auf die Konsummotive im Luxussegment kann festgestellt werden, dass grundsätzlich eine Unterteilung in den außengerichteten Luxuskonsum, den innengerichteten Luxuskonsum sowie den hier als hybrid bezeichneten Konsumtreibern erfolgt (Tab. 1).

Tab 1. Drei Kategorien für Motive im Luxuskonsum

Außengerichtete Konsummotive	Innengerichtete Motive	Hybride Motive
Demonstrativer Konsum/Signaling	Hedonismus/Genuss	Qualität
Soziale Schichtung	(Luxus-)Erleben	Nutzbarkeit
Status und Prestige	Selbstverwirklichung	Einzigartigkeit
Tradition	Erfüllung	Materialismus
Selbstdarstellung	Individualität	Preis
	Inspiration	
	Authentizität	

Ersterer basiert auf Konsummotiven, die erst innerhalb von Sozialgefügen Bedeutung erlangen. Sie setzen den Vergleich mit Mitmenschen voraus und beschreiben zunächst die Intention des *Signaling* von sozialer Stellung und Milieuzugehörigkeit (Han, Nunes, & Dreze, 2010;

Wiedmann, Hennigs, & Siebels, 2009). Hierbei findet eine enge Verknüpfung mit dem bei Veblen (1899) thematisierten *Demonstrativen Konsum* statt. Prestige, Status, Tradition und Selbstdarstellung sind weitere in der Literatur beschriebene außengerichtete Motive (Ascheberg, Meurer, & Oesterling, 2012; Vigneron & Johnson, 2004). Mason (1993) beschreibt drei verschiedene soziale Effekte, die durch den Konsum von Luxus erzeugt werden: den Veblen-, den Snob- und den Mitläufer-Effekt. Während der Veblen-Effekt eines Luxusgutes Demonstration von Reichtum erfasst, beschreibt der Snob-Effekt die Demonstration von Vornehmheit und sozialer Erhabenheit. Der Mitläufer-Effekt bezieht sich auf ein Unterstreichen von Zugehörigkeit zu einer bestimmten Bevölkerungsgruppe. Der Snob-Effekt verstärkt sich mit der Verringerung von Einkommensungleichheiten innerhalb der Gesellschaft, während der Veblen-Effekt in Zusammenhang mit einer Aufwärtsmobilität innerhalb der Gesellschaft steht. Er ist Ausdruck von Werten wie Erfolg, materialistischem Besitz und sozialem Aufstieg und beschreibt dabei die positive Korrelation zwischen der Nachfrage und dem Preis eines Gutes (Veblen, 1899).

Vigneron und Johnson (1999) erstellen ein Rahmenkonzept zum Verhalten von Konsumenten in Bezug auf Prestige-Güter und erweitern die zuvor identifizierten Konsummotive des Veblen-, Snob- und Mitläufer-Effektes um Hedonismus und Perfektionismus. Mit diesen fünf Motiven korrespondieren Werte, aus denen die Nachfrage nach Prestigegütern resultiert. Das Bedürfnis aufzufallen unterliegt demnach dem Veblen-Effekt, Einzigartigkeit ist in Verbindung zu bringen mit dem Snob-Effekt, der Mitläufer-Effekt resultiert aus einer starken Wahrnehmung des sozialen Umfelds des Konsumenten, bei den Hedonisten ist die emotionale Komponente vorherrschend und schlussendlich strebt der perfektionistische Käufer nach Maximierung des wahrgenommenen Nutzwertes. Die Autoren liefern damit ein mehrdimensionales Konzept und schaffen eine Verbindung zwischen den außengerichteten, den innengerichteten und den hybriden Motiven für Luxuskonsum.

Innengerichtete (Luxus-)Konsummotive resultieren aus dem Bedürfnis nach Genuss, Selbstverwirklichung und Erfüllung. Luxuskonsum dient hier der Verwirklichung hedonistischer Ziele sowie der Reifung und Auslebung von Persönlichkeit und Individualität (Meurer, 2012). Der Wert eines Luxusgutes ergibt sich aus innerpersönlichem Empfinden und emotionalem Zugewinn und ist damit losgelöst von der Eingliederung des Konsumenten in sein soziales Umfeld (Atwal & Williams, 2009). Das „Luxus-Erleben" sowie die Befriedigung subjektiver Ansprüche und Verlangen, wie das Streben nach Perfektionismus, sind Beispiele für die Thematisierung dieses Konzeptes in der Literatur (Atwal & Williams, 2009; Meurer, 2012, Vigneron & Johnson, 1999). Yeoman und McMahon-Beattie (2006) beschreiben einen Trend zu Luxuskonsum als Instrument für ein genussvolleres Leben, für die Erfüllung des Bedürfnisses nach Authentizität, Erleben und Hingabe. Tsai (2005) zeigt empirisch, dass ein persönlichkeitsbezogener Wert von Luxus die Kaufintention von Konsumenten für eine Luxusmarke signifikant erhöht. Kisabaka (2001) thematisiert den ausgeprägten Hang von Luxuskonsumenten zu Hedonismus und identifiziert u.a. das Bedürfnis zur Selbstentfaltung als maßgebliches Motiv beim Luxuskonsum. Ascheberg, Meurer, und Oesterling (2012) identifizieren einen Paradigmenwechsel „vom Haben zum Sein und von außen nach innen" im Luxuskonsum, der hedonistische Motive in den Vordergrund stellt. *Soziales Signaling* ist demnach nur noch ein Konsumziel neben anderen bei Luxusgütern.

Die dritte Kategorie der hybriden Konsumtreiber wird hier über die Merkmale Qualität, Nutzbarkeit, Einzigartigkeit, Materialismus und Preis definiert. Diese nehmen in Abhängigkeit vom einzelnen Konsumenten unterschiedliche Funktionen bei der Befriedigung außengerichteter oder innengerichteter Konsummotive ein. Demzufolge ist eine generelle Einordnung dieser Merkmale in die beiden vorangegangenen Kategorien nicht möglich. Eine gute Gebrauchstauglichkeit und Qualität, Einzigartigkeit sowie ein hoher materialistischer Wert werden als Charakteristika von Luxusprodukten bezeichnet (Hornig, Fischer, & Schollmeyer, 2013; Wiedmann, Hennigs, & Siebels, 2009), können aber sowohl als Instrument zum *Signaling* wie auch der Befriedigung persönlicher (hedonistischer) Bedürfnisse dienen (Vigneron & Johnson, 1999; Yeoman & McMahon-Beattie, 2006).

Der Preis eines Luxusgutes kann als Indikator für verschiedene Motive identifiziert werden. Insofern hat er eine mehrdimensionale Funktion. Hohe Preise werden in Zusammenhang gebracht mit hoher Qualität, Einzigartigkeit und Prestige (Erickson & Johansson, 1985; Kisabaka, 2001; Vigneron & Johnson, 2004; Wheatley & Chiu, 1977; Wiedmann, Hennigs, & Siebels, 2009). Als Charakteristikum von Luxusprodukten dienen Preise nicht zuletzt als Instrument zur Abgrenzung des Luxusmarktes von anderen Marktsegmenten (Serraf, 1991).

Einige Autoren thematisieren die daraus resultierende Maßgeblichkeit der Preispolitik in der Luxus-Vermarktung. Fassnacht, Kluge, und Mohr (2013) bestätigen die Relevanz des Veblen-Effektes und weisen auf die Vorteilhaftigkeit von luxusspezifischen Preismanagementprozessen hin. Hohe, konstante Preise steigerten das Luxusmarkenimage. Kapferer und Bastien (2009) empfehlen ebenfalls die systematische Anhebung von Preisen für Luxusgüter. Nicht nur werde durch hohe Preise die Aufmerksamkeit von Luxuskäufern geweckt, auch ließe sich durch sie ein ausgeprägtes Verantwortungsbewusstsein gegenüber dem Kunden in der Unternehmenskultur von Luxusunternehmen erzeugen. Darüber hinaus stellen Hornig, Fischer, und Schollmeyer (2013) fest, dass die Preise von Luxusgütern umso höher sind, je maskuliner oder individualistischer die Kultur eines Landes geprägt ist bzw. je größer Machtdistanzen gehalten werden.

Yeoman und McMahon-Beattie (2006) zufolge ist die Bereitschaft von Luxuskäufern zur Zahlung hoher Premium-Preise eng verknüpft mit dem empfundenen emotionalen Zugewinn, den das Produkt liefert. Luxus wird konsumiert, um dem (modernen) Bedürfnis nach Selbstverwirklichung, Genuss und Erleben nachzukommen. Damit rechtfertigt erst ein dem Produkt inhärentes hohes Potential zur Entsprechung dieser innengerichteten Konsummotive dessen Premium-Vermarktung. Auch Hagtvedt und Patrick (2009) stellen die Tauglichkeit zur Erfüllung hedonistischer Bedürfnisse als maßgebliches Merkmal heraus, das Luxusmarken von anderen unterscheidet. Dabei stoßen sie auf einen positiven Zusammenhang zwischen wahrgenommenem hedonistischen Potential und Preisniveau. Folglich findet eine Beeinflussung des innengerichteten, subjektiven Luxuswerts durch die Preispolitik einer Luxusmarke statt.

In einer Studie von Dubois, Czellar, und Laurent (2005) zur Konsumentensegmentierung auf Grundlage der Einstellung zu Luxus bewerten Probanden verschiedener Nationalitäten 33 Aussagen zu Luxus anhand einer fünfstufigen Likert-Skala. Die Autoren untersuchen 19 westlich geprägte Länder (darunter auch Deutschland) sowie die südasiatische Stadt Hong Kong und ordnen sie in einem dreidimensionalen Raum mit den Achsen *elitär*, *distanziert* und *demokratisch* an. Deutschland befand sich mittig im Dreieck, dementsprechend wurde es weder als Nation der starken Luxusaffinität noch als Nation der stark luxusaversen Menschen identifiziert. In diesem Fall scheint ein Ausgleich der Subkulturen stattzufinden.

Luxus in Deutschland wird ebenfalls bei Meurer (2012) thematisiert. Hier wird ein Wertewandel auf dem deutschen Luxusmarkt konstatiert, durch den innengerichtete Konsummotive wie Nachhaltigkeit und Verantwortung in den Vordergrund rücken. Der Konsum von Luxus findet so gesellschaftliche Rechtfertigung. Es entwickelt sich ein selbstverstärkender Prozess, der den aktuellen Trend zu einer größeren Nachfrage nach Luxus beschleunigt.

Reitsport und Luxus

Im Allgemeinen konnte mehrfach gezeigt werden, dass Reiter einkommensstark sind und ein hohes Bildungsniveau aufweisen (Tab. 2). Darüber hinaus zeichnen sie sich durch ein hohes Involvement in ihren Sport sowie hohe Zahlungsbereitschaften für Produkte und Dienstleistungen, bezogen auf Pferdehaltung und Reitsport, aus (Ikinger et al., 2013. Institut für Demoskopie Allensbach, 2013, IPSOS, 2003).

Sportwissenschaftliche Studien von Freyer (2001; 2003), Gille, Hoischen-Taubner und Spiller (2011), Häggblom, Rantamäki-Lahtinen und Vihinen (2012) und IPSOS (2003) ermittelten, wie oben zitiert, Reitsportmotive, welche sich z.T. einordnen lassen in die von Mason (1993) sowie Vigneron und Johnson (1999) vorgenommene Kategorisierung der Luxuseffekte. Einerseits kommt der Veblen-Effekt insofern zur Anwendung, als dass Erfolg und Anerkennung zu den Antriebskräften im Reitsport gehören. Weiterhin sind der Wunsch nach Geselligkeit und Gemeinschaft sowie die Aussicht, wichtige Menschen zu treffen, für den Reitsport relevante, sozialbezogene Motive. Andererseits konnte aber ein höherer Stellenwert der innergerichteten Motive identifiziert werden. Hedonismus in Form von Spaß haben, die Natur erleben, Freude am Pferd, Abenteuer, Erholung und Lust werden häufig als starke Motivatoren genannt.

Insgesamt geht aus der Literatur eine Bipolarität der Konsummotive sowohl im Luxusmarkt als auch im Reitsport hervor. Es kann eine Abgrenzung zwischen den außengerichteten und innengerichteten Motiven vorgenommen werden. Beide Formen treten parallel in Erscheinung, wobei letztere seit einigen Jahren deutlich an Bedeutung gewinnt. Demzufolge hat sich zwischenzeitlich eine Kategorisierung der innengerichteten Motive in den modernen Konsum und der außengerichteten Motive, insbesondere des demonstrativen Konsums, in den traditionellen Konsum etabliert.

In Anknüpfung an die Kategorisierung der Luxusmotive wurde die oben genannte Forschungsfrage dieser Studie entwickelt. Auf Grundlage der Ergebnisse werden Implikationen für das Luxusgütermarketing abgeleitet.

Methode

Im März 2013 wurden 646 Reitsportinteressierte mittels einer Online-Umfrage (615 Proban-den) und Face-to-Face-Interviews auf zwei Reitsportveranstaltungen (31 Probanden) zu ihren Einstellungen in Bezug auf Luxus befragt. Letztere dienten zur teilweisen Kompensation des für Online-Stichproben typisch hohen Anteils an jungen, noch in der Ausbildung befindlichen Probanden mit geringem Einkommen (Batinic & Bosnjak, 2000). Als Umfragemedium diente in beiden Fällen der gleiche standardisierte Fragebogen, dessen Beantwortung im Schnitt etwa 17 Minuten in Anspruch genommen hat. Da die Umfrage auf zwei deutschen Internetseiten (*http://www.st-georg.de* und *http://www.engarde.de*) verlinkt war und als Ort für die Face-to-Face-Interviews zwei deutsche Pferdesportveranstaltungen (in Essen und Mohnheim, beides Nordrhein-Westfalen) gewählt wurden, handelt es sich bei den Probanden ausschließlich um deutsche Pferdesportinteressierte. Obgleich durch die Beschränkungen des Samplings keine Repräsentativität der Studie gegeben ist, können auf Basis der Sondierungsstichprobe Rück-schlüsse auf die Forschungsfrage gezogen werden. Nicht die Größe der Segmente, sondern deren Charakteristika stehen im Fokus des Erkenntnisgewinns.

Die Konzipierung des Fragebogens orientierte sich an den Studien von Dubois und Laurent – (1994) und (1995) – und wurde dem Untersuchungsgegenstand entsprechend erweitert. Du-bois und Laurent (1995) entwickelten ein Instrument zur Messung der Luxusaffinität einzel-ner Konsumenten, die *Immersion scale*. Diese Skala umfasst acht luxuriöse Produkte sowie acht luxuriöse Aktivitäten. Ausgehend von der Anzahl der zutreffenden Items wird für jeden Probanden der Grad der Immersion in Luxus bestimmt. Aufgrund ihrer Eigenschaften, wie einer hohen Validität, Eindimensionalität, Objektivität sowie der Generierung eindeutiger Ergebnisse (Dubois & Laurent, 1995), kann die Skala zur wiederholten Anwendung empfoh-len werden. Die zweite Skala (Dubois & Laurent, 1994) umfasst 34 luxusbezogene, positive und negative Items. Es handelte sich dabei um Aussagen zur Einstellung zu Luxus im Allge-meinen, zum persönlichen Verhältnis zur Welt des Luxus sowie zur Einstellung gegenüber luxuskonsumierenden Mitmenschen. Auch diese Skala erweist sich als valide (Dubois & Lau-rent, 1994). Beide Messinstrumente sind an einer materiellen Luxusdefinition ausgerichtet. Insofern wird mit ihnen hauptsächlich Luxusaffinität in Bezug auf teure Güter und Luxus-marken eruiert (vgl. hierzu auch Hudders & Pandelaere, 2011).

Der Fragebogen zur vorliegenden Studie wurde in drei methodisch unterschiedliche Abschnit-te unterteilt. Der erste beinhaltete Fragen zur Rolle der Probanden im Pferdesport und diente überdies der Gewinnung sozio-demografischer Informationen. Der zweite Abschnitt bestand aus 40 Aussagen über Luxus im Allgemeinen, Luxusmarken und -produkte und Luxus in Verbindung mit dem Pferdesport, welche anhand einer fünfstufigen Likert-Skala zu bewerten waren (in Anlehnung an Dubois & Laurent, 1994). Der dritte Fragebogenteil bestand aus einer Liste mit 13 Luxusgegenständen und 12 luxuriösen Hobbys und Gewohnheiten (Erweiterung von Dubois & Laurent, 1995). Die Probanden wurden aufgefordert anzugeben, welche der Gegenstände sie besitzen bzw. welche der Aktivitäten sie in der angegebenen Regelmäßigkeit ausführen.

Die Auswertung erfolgte mittels zweier multivariater Verfahren. Zuerst wurden die Likert-skalierten Aussagen durch eine explorative Faktorenanalyse (Hauptkomponentenanalyse, Varimax-Rotationsmethode) zu vier Einstellungsdimensionen zusammengefasst. Anschließend diente eine dreistufige Clusteranalyse zur Aufspaltung der Stichprobe in vier Probandengruppen. Stichprobenausreißer wurden mittels des Single-Linkage-Verfahrens identifiziert und eliminiert. Anschließend konnte auf Grundlage der Ward-Methode eine sinnvolle Anzahl der Cluster bestimmt werden, woraufhin das K-Means-Verfahren zur Bestimmung der optimalen Endpartitionen diente (Hair et al., 1998). Vier in der Faktorenanalyse generierte Faktoren sowie die jeweiligen Häufigkeiten in Bezug auf die Listen der Luxus-Produkte und -Aktivitäten sind clusterbildende Variablen. Zur Clusterbeschreibung dienten acht Variablen.

Stichprobe

Wie schon vorherige, thematisch verwandte Studien ermittelten, stellte sich auch hier heraus, dass das Reiten ein frauendominierter Sport ist (Gille, Hoischen-Taubner & Spiller, 2011; IPSOS, 2003). Nur etwa 11% der Probanden waren Männer, wobei ihr Anteil in den Altersklassen ab 46 Jahren ansteigt, wie auch bei Gille, Hoischen-Taubner und Spiller (2011). Generell sind aber die 26- bis 45-Jährigen mit einem Anteil von 49,8% die am stärksten vertretene Gruppe.

In Bezug auf die Einkommensklassen fällt auf, dass der Anteil an Haushalten mit einem Nettoeinkommen von 3.600-5.000 Euro sowie einem Nettoeinkommen von 5.000 bis unter 18.000 Euro hoch ist (15,6% bzw. 13,2%). Zudem wurde ein hohes Bildungsniveau innerhalb der Stichprobe im Vergleich zur deutschen Bevölkerung vorgefunden. Die Anteile derer mit Abitur bzw. einem (Fach-) Hochschulabschluss als höchsten Bildungsabschluss liegen in der Stichprobe bei 36,8% und 37,6%. Über eine abgeschlossene Promotion verfügen 5,7% der Probanden. Für den Vergleich von sozio-demografischen Charakteristika der Stichprobe mit denen der Grundgesamtheit deutscher Reitsportler wird die Allensbacher Markt- und Werbeträgeranalyse herangezogen. In dieser wurden 25.677 deutsche Bundesbürger befragt. Unter anderem ermittelte man den Anteil der Reitsportler und dessen sozio-demografische Charakteristika (Institut für Demoskopie Allensbach, 2013; Rohdaten wurden den Autoren für diesen Zweck zur Verfügung gestellt). Daraus geht hervor, dass der Männeranteil der Grundgesamtheit der deutschen Reitsportler höher als in der Stichprobe ist. Reiten wird am häufigsten von unter 50-Jährigen ausgeführt (79,5% der Stichprobe), wobei die Altersklasse der 20- bis 29-Jährigen mit 22,4% den größten Anteil einnimmt. Weiterhin zeigte sich, dass 24,9% der deutschen Reiter über ein Nettohaushaltseinkommen von 2.500 Euro bis 3.500 Euro und 32,4% von 3.500 Euro oder mehr verfügen. Damit bezieht die Mehrheit ein Nettohaushaltseinkommen der beiden höchsten definierten Kategorien. Die Anteile der Abiturienten und Hochschulabsolventen fallen mit 21,7% und 17,2% im Vergleich zur Sondierungsstichprobe gering aus.

Tab 2. Prozentuale Häufigkeiten ausgewählter sozio-demografischer Charakteristika

Sozio-demografische Daten	Stichprobe	Grundgesamtheit der Reiter in DE[1]	Deutschland gesamt[2]
Geschlechterverteilung	Frauen: 89% Männer: 11%	Frauen: 74,1% Männer: 25,9%	Frauen: 51% Männer: 49%
Alter	Unter 15 Jahren: 0,3% 15-25 Jahre: 26,9% 26-45 Jahre: 49,8% 46-65 Jahre: 21,5% Über 65 Jahre: 0,6%[3]	14-19 Jahre: 20% 20-29 Jahre: 22,4% 30-39 Jahre: 18,6% 40-49 Jahre: 18,5% 50-49 Jahre: 11,7% 60-69 Jahre: 4,8% 70 Jahre und älter: 4%	14-19 Jahre: 6,9% 20-29 Jahre: 14% 30-39 Jahre: 13,6% 40-49 Jahre: 18,8% 50-49 Jahre: 16,8% 60-69 Jahre: 12,5% 70 Jahre und älter: 17,4%

Nettohaushalts-einkommen	Unter 1.300 Euro: 12,5% 1.300- unter 1.700 Euro: 10,8% 1.700- unter 2.600 Euro: 16,4% 2.600- unter 3.600 Euro: 13,3% 3.600- unter 5.000 Euro: 15,6% 5.000- unter 18.000 Euro: 13,2% 18.000 Euro oder mehr: 1,5%[4]	Unter 1.000 Euro: 6% 1.000- unter 1.500 Euro: 11,9% 1.500- unter 2.000 Euro: 12,9% 2.000- unter 2.500 Euro: 11,7% 2.500- unter 3.500 Euro: 24,9% 3.500 Euro und mehr: 32,6%	Unter 1.000 Euro: 6,9% 1.000- unter 1.500 Euro: 13,2% 1.500- unter 2.000 Euro: 16,5% 2.000- unter 2.500 Euro: 14,2% 2.500- unter 3.500 Euro: 25,2% 3.500 Euro und mehr: 24%
Höchster Bildungsab-schluss[5]	Abitur: 36,8% Universitäts- oder Fachhochschulabschluss: 37,6% Promotion: 5,7%[6,]	Abitur: 21,7% Universitäts- oder Fachhochschulabschluss: 17,2% Promotion: n. a.	Abitur: 10,9% Universitäts- oder Fachhochschulabschluss: 13,5% Promotion: 1,1%
Wohnort	Ländliche Region (unter 5.000 Einwohner): 34,4% Kleinstadt (5.000-100.000 Einwohner): 37,6% Großstadt (100.000-1.000.000 Einwohner): 19,6% Metropole (über 1.000.000 Einwohner): 8,4%	Unter 5.000 Einwohner: 15,7% 5.000- unter 20.000 Einwohner: 26,3% 20.000- unter 100.000 Einwohner: 29,4% 100.000- unter 500.000 Einwohner: 14,9% 500.000 und mehr Einwohner: 13,7%	Unter 5.000 Einwohner: 14,9%% 5.000- unter 20.000 Einwohner: 26,6% 20.000- unter 100.000 Einwohner: 27,2% 100.000- unter 500.000 Einwohner: 15% 500.000 und mehr Einwohner: 16,3%

[1]Quelle: Institut für Demoskopie Allensbach, 2013

[2]Quellen: Institut für Demoskopie Allensbach, 2013; Statistisches Bundesamt, 2013

[3]Keine Angabe machten 0,9%

[4]Keine Angabe machten 16,7%

[5]Nur die hier relevanten Werte

[6]Keine Angabe machten 0,9%

Ein Probandenanteil von 28% wohnt in einer Großstadt oder Metropole, während der Großteil in einer Kleinstadt bzw. einer ländlichen Region ansässig ist (72%). In der Grundgesamtheit beträgt der Anteil der Kleinstadt- und Landbewohner 71,4% (Institut für Demoskopie Allensbach, 2013). Tab. 2 zeigt die Häufigkeiten der hier relevanten sozio-demografischen Charakteristika für die Stichprobe, die Grundgesamtheit der Reiter in Deutschland sowie für Deutschland gesamt.

Das Involvement der Befragten in den Reitsport zeigte sich wie folgt: Etwa 50% der Probanden sind aktive Turnierreiter, während 31,8% ein eigenes und 54% sogar mehrere eigene Pferde besitzen. 59,8% der befragten Reiter widmen sich täglich ihrem Sport. Die Frage, wie viel Geld durchschnittlich im Monat für den Reitsport ausgegeben wird, war anhand von vorgegebenen Kategorien zu beantworten. Der höchste Anteil der Probanden (34,6%) gab an, 501-1.000 Euro monatlich für ihren Reitsport, inklusive der Kosten für eigene Pferde, auszugeben. Reiter, die im Dressursport aktiv sind, weisen im Übrigen die höchste Zahlungsbereitschaft für das Reiten auf. Zum Beispiel befinden sich unter ihnen etwa 70% derjenigen, die monatlich mehr als 2.000 Euro für ihren Sport investieren. Springreiter hingegen sind in dieser Kategorie nur mit 37,4% vertreten. Dressursport betreiben 76,3% der Probanden, damit ist diese Reitsportart die beliebteste in der Stichprobe Die zweithäufigste Sparte bildet das freizeitmäßige Reiten und Ausreiten (51,2%). Letzteres unterstreicht die Erkenntnisse aus der Werte- und Motivationsforschung im Bereich des Reitsports. Viele Reitsportler kommen offensichtlich ihrem Bedürfnis nach Entspannung, Spaß und der Gelegenheit, Natur zu erleben mittels Freizeitreiterei bzw. Ausreiten nach. Tabelle 3 zeigt die hier genannten Ergebnisse.

Tab 3. Häufigkeiten reitsportbezogener Variablen in Prozent

	Anzahl der Be- rufs-reiter	Anzahl der Turnier- reiter	Reiterliches Niveau	Häufigkeit der Ausübung des Reitsports	Anzahl der Reitsportler mit eige- nem/eige- nen Pferd(en)	Monatliche Ausgaben für den Reitsport
Grad der Involviertheit	6,5%	50,1%	Fortgeschritten (bis Klasse A): 17,8% Weit fortge- schritten (bis Klasse L): 39,8% Sehr weit fortgeschritten (bis Klasse S): 31,5%	Täglich: 59,8% Vier- bis sechsmal wöchentl.: 26,1% Ein- bis drei- mal wöchentl.: 10,7% Ein- bis zwei- mal monatl.: 1% Gelegentlich: 2,1% Nur im Urlaub: 0,3%	85,8%	Weniger als 50 Euro: 3% 50-100 Euro: 6,7% 101-500 Euro: 32,5% 501-1.000 Euro: 34,6% 1.001-1.500 Euro: 8,5% 1.501-2.000 Euro: 5% Mehr als 2.000 Euro: 4,7%[1]
Disziplinen[2] **(Anzahl der Aktiven in den jeweiligen Reitsportdisziplinen, Mehrfachantworten möglich)**	**Dressur**	**Springen**	**Vielseitigkeit**	**Fahren**	**Jagdreiten**	**Freizeitrei- ten/ Ausrei- ten**
	76,3%	37,4%	10,4%	3,4%	4%	51,2%

[1] Keine Angabe machten 5%.

[2] nur Disziplinen mit einer Häufigkeit $\geq$ 3%

Faktorenanalyse

Ziel der Studie war eine Clusteranalyse zur Spezifikation von Marketingzielgruppen innerhalb der Gruppe der Reitsportler. Als maßgebendes Kriterium diente der Grad der Luxusaffinität. Anhand von sechs clusterbildenden Variablen wurde eine Skala zur Messung der Einstellungen gegenüber Luxus entwickelt und damit eine Grundlage für die Identifizierung heterogener Konsumentengruppen geschaffen. Die Struktur der mit der Varimax-Methode ermittelten Faktoren sowie die Verteilungsparameter der faktorenbildenden Items sind in Tabelle 4 dargestellt. Es wird deutlich, dass die Bewertung der Aussagen annähernd einer Normalverteilung folgt. Als Grundlage für die Faktorenbildung dienten die oben bereits erwähnten, an Dubois und Laurent (1994), Dubois, Czellar und Laurent (2005) angelehnten 40 Aussagen zu Luxus und neu formulierte reitsportspezifische Statements. In der endgültigen Lösung wurden die 18 Variablen mit dem höchsten Erklärungsgehalt berücksichtigt.

Tab 4. Einstellungsdimensionen zum Luxus im Reitsport[1]

Faktor 1: Affinität zu Luxusprodukten (Cronbachs Alpha: 0,78, erklärter Anteil der Gesamtvarianz: 21,12%)	Ladung	μ ; σ
Ich informiere mich regelmäßig über die neuen Trends in Mode, Kosmetik, Einrichtungen und Lifestyle.	0,78	-0,44 ; 1,10
Ich mag es, mich mit Freunden über neu erworbene Gegenstände, Mode, Design und Lifestyle zu unterhalten.	0,68	-0,47 ; 1,06
Ich kenne mich mit Luxus aus.	0,67	-0,22 ; 0,90
Ich kaufe oft Luxusprodukte.	0,63	-0,01 ; 1,00
Grundsätzlich mag ich Luxus.	0,63	0,47 ; 0,79
Wenn ich mich in einem Geschäft für Designerprodukte aufhalte, habe ich besondere Glücksgefühle.	0,61	0,77 ; 1,09
Faktor 2: Ablehnende Haltung gegenüber Luxusprodukten und deren Käuferschaft (Cronbachs Alpha: 0,7, erklärter Anteil an der Gesamtvarianz: 18,91%)	**Ladung**	**μ ; σ**
Menschen, die Luxusprodukte kaufen, wollen damit die Reichen imitieren.	0,76	-0,18 ; 0,88
Ich assoziiere negative Attribute mit der Persönlichkeit eines Menschen, der teure Luxusprodukte kauft.	0,72	-0,52 ; 0,87
Für die meisten Menschen dienen Luxusprodukte nur als Geschenke, die sie für andere kaufen oder die sie sich von anderen wünschen.	0,59	-0,45 ; 0,77
Man muss ein bisschen abgehoben sein, um teure Marken- oder Designerwaren zu kaufen.	0,58	-0,30 ; 0,98
Luxus sollte stärker besteuert werden.	0,52	-0,43 ; 1,10
Menschen, die hochwertige Produkte kaufen, wollen sich damit von anderen abheben.	0,52	0,16 ; 0,92
Faktor 3: Snob-Effekte (Cronbachs Alpha: 0,64, erklärter Anteil der Gesamtvarianz: 10,77%)	**Ladung**	**μ ; σ**
Wenn ich hochwertige Produkte konsumiere oder ein kostspieliges Hobby betreibe, stört es mich, wenn ich von vielen Menschen umgeben bin bzw. Rücksicht auf andere nehmen muss.	0,85	-0,51 ; 0,97
Wenn ich für ein Hobby viel Geld investiere, möchte ich mich dabei nur in einer kleinen Gruppe Gleichgesinnter aufhalten.	0,82	-0,52 ; 0,98

Faktor 4: Exklusiver Reitsport ist Luxus (Cronbachs Alpha: 0,78, erklärter Anteil der Gesamtvarianz: 7,46%)	Ladung	μ ; σ
Für mich ist es Luxus, mein Pferd in einem teuren Stall mit tollen Bedingungen unterbringen zu können.	0,81	0,36 ; 1,19
Für mich ist es Luxus, ein teures Pferd zu besitzen.	0,81	0,25 ; 1,24
Für mich ist es Luxus, zum Kreis der Akteure des kostspieligen Reitsports zu gehören.	0,71	-0,38 ; 1,15
Für mich ist es Luxus, mir eine/n gute/n Ausbilder/in für mein Pferd leisten zu können.	0,70	0,59 ; 1,07

[1] Aussagen waren anhand von fünfstufiger Likert-Skala (von 0% Zustimmung bis 100% Zustimmung) zu beantworten; Faktorenanalyse auf Basis von Varimax-Rotationsmethode mit Kaiser-Normalisierung; 5 Iterationen, Kayser-Meyer-Olkin-Kriterium: 0,83; Signifikanz der Werte nach Bartlett: 0,000; erklärter Anteil der Gesamtvarianz: 58,26%

Faktor 4 bezieht sich auf die Wahrnehmung von Luxus, verknüpft mit dem Reitsport, die drei anderen Faktoren hingegen korrespondieren mit der Einstellung gegenüber Luxus- und Designer- bzw. Markenprodukten. Alle vier Faktoren weisen mit Werten um 0,7 ein gutes Cronbachs Alpha auf (Flynn, Schroeder, & Sakakibara, 1994). Nicht aussagekräftige bzw. nur mit geringen Ladungen versehene Faktoren erfuhren in der Clusteranalyse keine Berücksichtigung. Zusätzlich zu den vier Einstellungsdimensionen wurden zwei weitere verhaltensorientierte clusterbildende Variablen bestimmt. Die Addition der Luxusaktivitäten und -güter ergab per Indexbildung je Proband zwei Variablen – die Summe der zutreffenden Aktivitäten sowie die Summe der Luxusgüter. Anhand der von Dubois und Laurent (1995) entwickelten *Immersion scale* können Aussagen zur Luxusaffinität der Konsumentengruppe getroffen werden. In dieser Studie wurde der jeweilige Katalog erweitert und an die Thematik angepasst. Bei der deskriptiven Auswertung der Skala zeigt sich, dass in dieser Stichprobe luxuriöse Hobbys und Aktivitäten tendenziell häufiger ausgeführt werden als dass luxuriöse Güter besessen werden (Tab. 5).

Tab 5. Katalog der luxuriösen Hobbys/Aktivitäten und Produkte, angelehnt an Dubois und Laurent (1995)

Luxuriöse Hobbys/Aktivitäten und die Regelmäßigkeit ihrer Ausführung	Ausführungs-häufigkeit
Wochenendtrip auf eigene Kosten während des letzten Jahres	50,2%
Restaurantbesuch auf eigene Kosten im Wert von 50 Euro oder mehr pro Person während der letzten zwei Monate	33,9%
Wöchentliche Einkäufe in Bio- oder Delikatessenläden	20,8%
Überseeflug auf eigene Kosten während des letzten Jahres	14,5%
Regelmäßiger Genuss (mind. 1x pro Woche) von Premiumweinen/-spirituosen oder Champagner	12,5%
Wöchentlicher Reit-Einzelunterricht mit einem Preis pro Unterrichtseinheit von 60 Euro oder mehr	11,2%
Regelmäßiges Ausführen (mind. 6x pro Jahr) von Freizeitaktivitäten wie Golfen, Segeln oder Skifahren	9,2%
Außerschulisches Erlernen einer Fremdsprache	9%
Außerschulisches Erlernen eines Musikinstruments	7,4%
Regelmäßiger Besuch (mind. 6x pro Jahr) von Wellnesseinrichtungen oder Kosmetikinstituten der gehobenen Preiskategorie	6,9%
Mitgliedschaft in einem Premium-Fitnessstudio	3,2%
Laufende Spekulation an der Börse	3,2%
Luxuriöse Güter	Besitzhäufig-keit
Kleidungsstück im Wert von 500 Euro oder mehr	27,2%
Immobilie mit einer Größe von 200qm oder größer als Erstwohnsitz	23,7%
Schmuckstück im Wert von 1.000 Euro oder mehr	19%
Armbanduhr im Wert von 1.000 Euro oder mehr	12,9%
Käuflich erworbenes Kunstwerk im Original	10,2%
Auto im Wert von 60.000 Euro oder mehr	9,7%
Gepäckstück (Koffer, Handtasche) im Wert von 500 Euro oder mehr	9,1%
Silberbesteck, welches nicht geerbt wurde	8,9%
Kosmetikprodukt im Wert von 200 Euro oder mehr	7,3%
Pferde-Transporter mit Wohnbereich (Schlafplätze, Nasszelle und Kochmöglichkeiten)	4,2%
Urlaubsresidenz/Wochenendhaus	4,2%
Porzellanservice im Wert von 3.000 Euro oder mehr, welches nicht geerbt wurde	3,6%
Motor- oder Segelyacht	0,9%

Bei einer Clusteranalyse konnten auf Basis der im Methodik-Teil beschriebenen dreistufigen Vorgehensweise die im Folgenden dargestellten vier Konsumentensegmente ermittelt werden.

Clusteranalyse

Die Güte der Clusteranalyse wurde mittels Post-hoc-Analyse und Kanonischer Diskriminanzfunktion bestimmt. Erstere bestätigte signifikante Unterschiede zwischen den einzelnen Gruppen auf dem 5%-Niveau (Tab. 6). Der dritte Faktor zeigt sich als am wenigsten aussagekräftig hinsichtlich der Trennung der Segmente, während Faktor 1 sowie die an Dubois und Laurent (1995) angelehnte, erweiterte *Immersion scale* („Anzahl der Luxusprodukte" und „Anzahl der luxuriösen Hobbys/Aktivitäten") die größte Trennkraft aufweisen. Die Mittelwerte der Segmente 1 und 2 unterscheiden sich mit Ausnahme des Faktors 3 hoch oder höchst signifikant. Segment 1 und 3 unterscheiden sich in vier der sechs clusterbeschreibenden Faktoren hoch oder höchst signifikant, dort jedoch mit einer höheren Ausprägung der Mittelwertdifferenzen als die Erstgenannten. Die geringsten Mittelwertabweichungen ergeben sich hingegen für Segment 1 und 4. Die Abgrenzung dieser beiden Cluster geschieht hauptsächlich über Faktor 1 (Luxusaffinität) und die Ergebnisse aus der erweiterten *Immersion scale*. Für die Abgrenzung der Segmente 2 und 3 lässt sich feststellen, dass die vier Faktoren im Vergleich zur erweiterten *Immersion scale* die höchsten Unterschiede bilden.

Die Diskriminanzanalyse klassifizierte 97,4% der ursprünglich gruppierten Fälle korrekt. Für ein Niveau von 0,05 wurden hochsignifikante Koeffizienten generiert. Der Test der Funktionen eins bis drei ermittelte ein Wilks-Lambda von 0,067, was zeigt, dass nur 6,7% der Streuung der aktiven Variablen nicht durch die Gruppenunterschiede erklärbar sind. Die Werte für den Test der Funktionen zwei bis drei und drei wiesen mit 0,387 und 0,711 schlechtere Ergebnisse für dieses Gütemaß aus. Dennoch führte die Clusteranalyse insgesamt zu aussagekräftigen und plausiblen Ergebnissen.

Tab 6. Ergebnisse der Post-hoc-Analyse: Mittlere Differenzen aus Mehrfachvergleichen

Cluster		Fakt. 1: Affinität	Fakt. 2: Ablehnung	Fakt. 3: Snob-Effekte	Fakt. 4: Exklusiver Reitsport	Anzahl der Luxusprodukte in Besitz	Anzahl der luxuriösen Hobbys / Gewohnheiten
1	2	0,75167982***	-1,30***	-,17885203	-0,38**	4,96348***	3,04137***
	3	1,65251543***	-,15899912	,23579094	0,39012998**	4,96786***	3,1***
	4	0,54960176***	-,06140370	,09594781	-,18430123	4,02857***	0,88571***
2	1	-0,75***	1,29594374***	,17885203	0,3766173**	-4,96***	-3,04***
	3	0,90083561***	1,13694463***	0,41464297***	0,76674728***	,00438	,05863
	4	-,20207806	1,23454004***	0,2747998386*	,19231607	-0,93***	-2,16***
3	1	-1,65***	,15899912	-,23579094	-0,39**	-4,97***	-3,10***
	2	-0,90***	-1,14***	-0,41464297***	-0,77***	-,00438	-,05863
	4	-1,10***	,09759542	-,13984313	-0,57443121***	-0,94***	-2,21***
4	1	-0,55***	,06140370	-,09594781	,18430123	-4,03***	-0,89***
	2	,20207806	-1,23***	-,27479984*	-,19231607	0,93491***	2,15566***
	3	1,10291367***	-,09759542	,13984313	0,57***	0,93929***	2,21429***

* signifikant auf dem Niveau 0,1 ; **signifikant auf dem Niveau 0,05 ; ***signifikant auf dem Niveau 0,01

Anhand der clusterbildenden und -beschreibenden Variablen können die Reitersegmente wie folgt beschrieben werden (Tab. 7):

Eine relativ kleine Gruppe von siebzig der insgesamt 646 Probanden (etwa 11%) konnte dem Profil des „luxusaffinen Reitsportlers" mit überdurchschnittlich hoher Bildung und hohem Einkommen zugeordnet werden. Die positiv formulierten Aussagen zu Luxus wurden überwiegend mit „Stimme zu" beantwortet, während die negativ formulierten Aussagen tendenziell auf Ablehnung stießen. Die Anzahl der einschlägigen Hobbys und Güter waren mit vier und sechs hoch im Vergleich zu den anderen drei Segmenten. Geld für Reitsport auszugeben, wurde in der Eigenwahrnehmung tendenziell nicht als Luxus empfunden. Dies könnte allerdings dem Umstand geschuldet sein, dass diese Gruppe über ein überdurchschnittliches Einkommen verfügt – die am häufigsten angekreuzte Einkommenskategorie ist 5.000-10.000 Euro Nettohaushaltseinkommen – und so Investitionen in den Reitsport als selbstverständlich wahrgenommen werden. Der Anteil der auf Turnieren aktiven Reiter von 51,4% sowie der

Anteil der Pferdebesitzer von über 97,1% weist auf einen hohen Grad an Involvement in den Reitsport hin. Insgesamt handelt es sich bei den Konsumenten des Clusters 1 um erfolgreiche, sehr gut verdienende Menschen mittleren Alters, die Luxus konsumieren und genießen und überdies eine hohe Zahlungsbereitschaft für den Reitsport aufweisen (die durchschnittlichen monatlichen Ausgaben für den Reitsport liegen bei 1.400 Euro).

Etwa 33% der befragten Reitsportler (212 Probanden) gehören zur zweiten Gruppe, „die Luxus-Ambivalenten", und können ebenfalls als ambitionierte Reiter charakterisiert werden. Wiederum treten 51,4% der Gruppe auf Turnieren an und 82,1% besitzen ein oder mehrere Pferde. Im Gegensatz zur ersten Gruppe üben diese Reiter allerdings keinen luxuriösen Lebensstil aus, was in Einklang mit der am häufigsten vorgefundenen Einkommenskategorie von 1.300-1.700 Euro steht. Allerdings wurden auch hier die positiv formulierten Statements zu Luxus eher zustimmend bewertet. Auffällig an dieser Gruppe ist jedoch, dass ebenfalls sowohl den negativ formulierten Aussagen sowie den Items zu Snob-Effekten leicht zugestimmt wurde. Zugleich wird der exklusiv betriebene Reitsport (teurer Stall und gute Ausbildung von Pferd und Reiter) tendenziell als Luxus empfunden. Dementsprechend kann davon ausgegangen werden, dass die Konsumenten dieser Gruppe sich grundsätzlich nicht in „der Welt des Luxus" wiederfinden, sich aber trotzdem mit dem Thema beschäftigen. Sie weisen eine widersprüchliche Einstellung gegenüber Luxus und denen, die ihn konsumieren, auf. In Relation zu ihrem Einkommen weist diese Gruppe eine hohe Zahlungsbereitschaft für den Reitsport auf (monatliche Ausgaben von gerundet 500 Euro). In der Tendenz ist diese Gruppe zehn Jahre jünger als die vorherige und zu 40,9% in einem Angestelltenverhältnis beschäftigt. Der Reitsport ist ihnen wichtig, dafür geben sie einen Großteil ihres Gehalts aus.

Tab 7. Gruppierungen der Pferdesportler in Deutschland anhand des Kriteriums der Luxusaffinität[5]

Clusterbildende Variablen	Mittelwerte			
	Segment 1: Die Luxusaffinen (N=70)	Segment 2: Die Luxus-Ambivalenten (N=212)	Segment 3: Die Luxus-Abgeneigten (N=224)	Segment 4: Die Reitsport-fokussierten (N=140)
Affinität zu Luxusprodukten*	0,934	0,183	-0,718	0,385
Ablehnende Haltung gegenüber Luxusprodukten und deren Käuferschaft*	-0,491	0,805	-0,332	-0,429
Snob-Effekte*	0,041	0,22	-0,195	-0,055
Exklusiver Reitsport ist Luxus*	-0,033	0,344	-0,423	0,151
Anzahl der luxuriösen Hobbys und Gewohnheiten*	4	1	1	4
Anzahl der luxuriösen Produkte und Güter in Besitz*	6	1	1	2
Clusterbeschreibende Variablen	Segment 1: Die Luxusaffinen (N=70)	Segment 2: Die Luxus-Ambivalenten (N=212)	Segment 3: Die Luxus-Abgeneigten (N=224)	Segment 4: Die Reitsport-fokussierten (N=140)
Altersmedian der Gruppenmitglieder	40 Jahre	31 Jahre	40 Jahre	37 Jahre
Häufigste Einkommensgruppe der Gruppenmitglieder (gemessen am monatlichen Nettohaushaltseinkommen)	5000-10.000€ (24,2%)	1300-1700€ (17%)	1700-2600€ (20%)	3600-5000€ (22,1%)
Häufigster höchster Bildungsabschluss der Gruppenmitglieder	Universitäts-/Fachhochschulabschluss (43,3%), höchste Promoviertenrate (11,9%)	Abitur/Fachabitur (43,8%)	Abitur/Fachabitur (38,3%)	Universitäts-/Fachhochschulabschluss (51,4%)

---

[5] Der Anteil der Probanden aus den Face-to-Face-Interviews beträgt im ersten Segment 8,6%, im zweiten 0,9%, im dritten 4% und im vierten 5%, d. h., ihr Anteil bei den Luxusaffinen ist der größte. Damit korrespondierend sind in dieser Stichprobe die Anteile der unteren Alters- und Einkommensklassen im Vergleich zur Online-Stichprobe unterrepräsentiert. Der Anteil der Probanden, die sich noch in Ausbildung befanden, unterliegt dem der Online-Stichprobe um etwa 5%. Diese Befunde entsprechen den Zielen, die zur Ergänzung der Online-Stichprobe durch eine persönliche Befragung geführt haben (siehe Methode).

Häufigster Berufsstand der Gruppenmitglieder	Selbstständige/r mit Angestellten (23,9%)	Angestellte/r (40,9%)	Angestellte/r (51,4%)	Angestellte/r (25%)
Durchschnittliche monatliche Zahlungsbereitschaft der Gruppenmitglieder in Bezug auf den Reitsport	1400€	500€	400€	800€
Turnierreiter (prozentualer Anteil der Turnierreiter innerhalb der Gruppe)	51,4%	51,4%	46,9%	53,6%
Pferdebesitz (prozentualer Anteil der Pferdebesitzer innerhalb der Gruppe)	97,1%	82,1%	85,7%	85,7%

*hoch signifikant (p ≤ 0,05)

Der dritten Gruppe der „Luxus-Abgeneigten" sind 35% der Probanden (N=224) zuzuordnen, wobei es sich wiederum um engagierte Reiter ohne luxuriösen Lebensstil handelt. Der Anteil der Turnierteilnehmer ist mit 46,9% der geringste, der Großteil (85,7%) besitzt aber ein oder mehrere eigene Pferde. Allerdings kann hier von einer gleichgültigen bis negativen Haltung zu Luxus und Luxuskonsumenten ausgegangen werden. Weiterhin konsumiert diese Gruppe nur wenig Luxus (tendenziell wird nur ein Luxusprodukt besessen bzw. nur eine Luxusaktivität regelmäßig ausgeführt) und zeigt zudem eine niedrige Zahlungsbereitschaft für den Reitsport in Relation zu den anderen Gruppen (durchschnittlich 400 Euro im Monat). Die häufigste Einkommenskategorie ist etwas höher als in Segment 2 (1.700-2.600 Euro) bei einem Altersmedian gleich dem der Gruppe 1 (40 Jahre).

Der vierten Gruppe, den luxusaffinen, ambitionierten Reitern mit einem besonderen Fokus ihrer Luxuspräferenzen auf den Reitsport, gehören 140 Reiter der Stichprobe (etwa 21%) an. Sie werden als „die Reitsportfokussierten" charakterisiert. Positiv formulierte Items zu Luxus wurden leicht zustimmend bewertet, negative Aussagen stießen auf Ablehnung. Ebenfalls ablehnend bewertet wurden Aussagen, die dem Faktor 3 (Snob-Effekte) zugeordnet sind (Tab. 4). Hieraus ist zu schließen, dass diese Gruppe sich gerne in Gemeinschaft anderer begibt und sich während des Konsums von Luxus nicht durch andere gestört fühlt. Die Anzahl der Luxus-Aktivitäten und -güter liegt bei vier bzw. zwei und für die Zahlungsbereitschaft für den Reitsport wurde bei dieser Gruppe der zweithöchste Wert ermittelt (800 Euro). Sich einen guten Stall und exzellente Ausbilder für Reiter und Pferd leisten zu können, wird, wie auch in Segment 2, als Luxus empfunden. Bei dieser Gruppe, die im Übrigen den höchsten Anteil an Turnierreitern aufweist (53,6%), nimmt der Reitsport einen hohen Stellenwert ein und Reitsport wird mit Luxus in Verbindung gebracht. Dazu passt, dass sich in dieser Gruppe die meisten Besitzer eines Pferde-Transporters, ausgestattet mit Schlafplatz, Nasszelle und Kochmöglichkeit, befinden (9 Probanden) sowie der höchste Anteil derer, die 60 Euro oder mehr für wöchentlichen Reit-Einzelunterricht ausgeben (33 Probanden). Das Nettohaushaltseinkommen dieser Gruppe ist zu einem Anteil von 22,1% in der Kategorie 3.600-5.000 Euro angesiedelt und damit überdurchschnittlich hoch. Der Altersmedian liegt bei 37 Jahren.

Diskussion

Die Faktorenanalyse formierte vier Einstellungsdimensionen gegenüber Luxuskonsum, welche zum Teil eng mit den vorangestellten, in der Literatur identifizierten Konsummotiven korrespondieren. Zwei Faktoren werden dominiert von dem bei Mason (1993), Vigneron und Johnson (1999) identifizierten Snob- bzw. Mitläufer-Motiv. Faktor 2 drückt eine negative Haltung gegenüber denjenigen aus, die Snob- oder Mitläufer-motivierten Luxuskonsum praktizieren. Indes transferiert Faktor 3 beide Konzepte in positiver Haltung.

In Anlehnung an das bei Wiedmann, Hennigs und Siebels (2009) beschriebene Segment der Materialisten, kann Faktor 1 (luxusaffine Haltung) als besitzorientiertes Motiv interpretiert werden. Die entsprechenden Statements drücken Befriedigung durch Kauf und Trendbewusstsein aus, sind aber persönlichkeitsbezogen und nicht vom sozialen Kontext abhängig. Damit stellt Faktor 1 auch die bei Kisabaka (2001) thematisierte Selbstdefinition und -belohnung durch materialistischen Konsum dar.

Faktor 4 beinhaltet Statements zum Konsum von Reitsport(-dienstleistungen), in welchen vorrangig den hybriden Motiven Qualität und Preis entsprochen wird. Luxus wird hier in Verbindung gebracht mit der Wahl eines teuren, gut ausgestatteten Stalls, eines guten Ausbilders, dem Besitz eines teuren Pferdes sowie der Tatsache, generell zu den Akteuren eines so kostspieligen Sports zu gehören. Die enge Verknüpfung von Luxus, Preis und Qualität in diesem Faktor lässt vermuten, dass auch hier der Preis als Indikator für gute Qualität wirkt (in Übereinstimmung mit Erickson & Johansson, 1985). Einen weiteren Anhaltspunkt für diese Annahme liefert der vergleichsweise hohe positive Mittelwert des Faktors im Cluster der Luxus-Ambivalenten. Letztere lehnen das Snob-Motiv vergleichsweise stark ab (Mittelwert für Faktor 2: 0,805). Deshalb ist nicht davon auszugehen, dass ihre Zustimmung zu den Statements des Faktors 4 auf einer Verknüpfung von hohen Preisen mit Prestige oder sozialer

Schichtung basiert. Wir erkennen hier eher eine Fokussierung der in Kapitel 2 dargestellten hybriden Konsummotive.

In der Clusteranalyse ergab sich für ein Drittel der Befragten (Segmente 1 und 4) ein positiver Mittelwert für Faktor 1 (Affinität zu Luxusprodukten) und ein negativer Mittelwert für Faktor 2 (ablehnende Haltung gegenüber Luxusprodukten und deren Käuferschaft). Das Drittel der Ambivalenten weist nur positive Mittelwerte auf, während im Segment der Abgeneigten Luxus ausschließlich negativ bewertet wird.

Die ambivalente Einstellung des zweiten Segments könnte in Zusammenhang mit dem geringen Durchschnittsalter von 31 Jahren und den überwiegend niedrigen Einkommen stehen. Es ist möglich, dass die Probanden dieses Segments im weiteren Verlauf ihres Lebens eine eindeutige, entweder affine oder averse Einstellung zu materiellem Luxus entwickeln. Die luxusaffinen und die reitsportfokussierten Befragten (Segmente 1 und 4) unterscheiden sich dahingehend, dass in Letzterem insbesondere der Konsum qualitativ hochwertiger Reitsportdienstleistungen als Luxus empfunden wird. Dort finden wir ein niedrigeres Einkommen und geringere Zahlungsbereitschaften als in Segment 1 vor, gleichzeitig aber eine vermehrte Zustimmung zu den im Faktor „Exklusiver Reitsport ist Luxus" (Faktor 4) zusammengefassten Statements. Eine ausgeprägte Zustimmung zum vierten Faktor beobachten wir ebenfalls im Segment der Ambivalenten. Dies lässt die Vermutung zu, dass insbesondere unter den Ambivalenten und den Reitsportfokussierten die im Reitsport dominanten, innengerichteten Motive wie Hedonismus, Gesundheit und Erlebnis zum Tragen kommen.

Generell wurde deutlich, dass Reiter ihr Hobby hoch engagiert ausführen und dass Reiten mehr ist als nur ein Ausgleich zum Alltag oder ein Mittel zur Erlangung körperlicher Fitness. Dies steht in Einklang mit vorangegangener Literatur (Ikinger et al., 2013). Das hohe zeitliche und emotionale Involvement der Akteure im Reitsport indiziert, dass diese Zielgruppe grundsätzlich mit Marketinginstrumenten, die das Pferd und den Reitsport thematisieren, erreicht werden kann. Der Kunde kauft eine Marke dann, wenn er sich mit ihr identifizieren kann. Er drückt durch die Markenwahl sein Selbstkonzept aus (Fournier, 1998; Strebinger, 2001). Dies wird auch für den Fall von Luxusmarken bestätigt (Wieseke et al., 2013). Die Verbundenheit der Reiter zu ihrem Sport schafft für Luxusmarken eine günstige Voraussetzung, identitätsstiftende Wirkung zu entfalten. Vorteilhaft für die Vermarktung von Luxusmarken ist auch das tendenziell hohe Einkommen von Reitsportlern. Den Reitsport, insbesondere die leistungsbezogene Ausführung, kennzeichnen besonders hohe Kosten (Hoogenraad & Vering, 2007). Dadurch wird den einkommensstarken Pferdeliebhabern im Vergleich zu einkommensschwachen der Zugang erleichtert, abermals bestätigt durch die Befunde in dieser Stichprobe.

Wie sich jedoch herausstellte, ist die Positionierung von Luxusmarketing, welches vornehmlich auf Luxusdimensionen wie Materialismus, Exklusivität und Prestige abstellt, im Reitsport nur dann sinnvoll, wenn zielgerichtet das in der Analyse identifizierte Segment der Luxusaffinen bedient werden soll. Das Segment der Reitsportfokussierten zeigte sich zwar luxusaffin, gleichzeitig wurden aber hier die Statements des dritten Faktors (Snob-Effekte) tendenziell abgelehnt. In Kombination mit der positiven Haltung dieser Gruppe gegenüber den Statements zu „Exklusiver Reitsport ist Luxus" weist dies auf ein Luxusverständnis hin, das sich von Materialität und sozialer Distinktion distanziert. Aufgrund positiver Mittelwerte für den ersten Faktor (Affinität zu Luxusprodukten) und hohen Zahlungsbereitschaften in beiden Segmenten kann aber unterstellt werden, dass die Luxusaffinen und Reitsportfokussierten grundsätzlich zur Zielgruppe von klassischen Luxusmarken gehören.

Segment 2 fällt aufgrund der niedrigen Einkommensklasse nicht in die Zielgruppe teurer Luxusgüter, und Segment 3 weist einen verhältnismäßig ausgeprägten negativen Mittelwert für den ersten Faktor (Affinität zu Luxusprodukten) auf. Hier stößt der materielle Luxusbegriff auf Ablehnung.

Der Reitsport und die Pferdehaltung sind Teil eines speziellen Lebensstilkonzepts. Mit Prestige und Distinktion wird es offensichtlich in einem vergleichsweise nur kleinen Marktsegment assoziiert. Vielmehr kann der Reitsport als *luxury experience* charakterisiert werden. Es zeigen sich die für dieses Konzept bekannten innengerichteten und immateriellen Konsummotive. Der Konsum von *luxury experience* dient dazu, den eigenen Lebensstil durch luxuriöses, sinnstiftendes Erleben zu bereichern (Atwal & Williams, 2009). Kommentare von Probanden, für die Platz am Ende des Fragebogens eingeräumt war, ließen darauf schließen, dass Zeit für das Pferd und den Reitsport als „wahrer Luxus" empfunden wird. So bezogen sich etwa 50 von insgesamt 87 Kommentaren auf eine persönliche Luxusdefinition, die hauptsächlich immaterielle Werte, wie Zeit, Gesundheit, Leben in Einklang mit der Natur und Genuss, umfasst. Zudem schrieben einige Geringverdiener, dass sie auf vieles, insbesondere auf teure Luxusartikel, verzichten, um reiten gehen zu können. Diese Kundensegmente stehen den einkommensstarken und luxusaffinen Probanden aus Segment 1 gegenüber, sollten aber dennoch in den Marketingerwägungen für Reitsportprodukte Beachtung finden. Beispielsweise zeigt sich in Studien, dass die Gruppe der jungen, noch in der Ausbildung befindlichen Menschen mit geringem Einkommen einen nicht vernachlässigbaren Anteil unter den Reitern ausmacht (Institut für Demoskopie Allensbach, 2013; IPSOS, 2003). Zudem findet sich gerade hier eine große Übereinstimmung mit dem vorher dargestellten Motivwandel und der damit einhergehenden wachsenden Relevanz immaterieller Werte, der sich sowohl im Luxuskonsum als auch im Reitsport vollzieht. Demnach bietet es sich an, Reiten als Lebensstil aufzugreifen und Luxus als immateriellen Wert bzw. Reiten an sich zu definieren. In diese Erwägung mit einzubeziehen ist auch die Häufigkeitsverteilung hinsichtlich der Wohnorte. Die Tatsache, dass mit etwa 70% der Großteil der Probanden in einer Kleinstadt oder einer ländlichen Region lebt, spricht dafür, dass die Wertschätzung für Naturerleben und einen ruhigen Lebensrhythmus kennzeichnend ist für die Konsumentengruppe der Reiter. Letzteres ergeben auch die Analysen bei Ikinger et al. (2013).

Insgesamt konnte festgestellt werden, dass in dieser Stichprobe die innengerichteten Motive maßgeblich sind bei der Erklärung des Konsums von Reitsport und Luxus. Die außengerichteten Motive treten nur vergleichsweise selten in Erscheinung, erzeugen segmentweise sogar Ablehnung. Insbesondere persönliches Erleben, Selbstverwirklichung und Genuss von immateriellem Luxus werden als Motivatoren genannt. Dies deckt sich mit den Befunden aus der sportwissenschaftlichen Literatur (D. R. Vereinigung, 2001; Gille, Hoischen-Taubner, & Spiller, 2011; Häggblom, Rantamäki-Lahtinen, & Vihinen, 2012). Gleichzeitig steht dieses Ergebnis in Einklang mit der im Luxusmarketing identifizierten aktuellen Relevanz von individuellem und persönlichkeitsorientiertem Konsum (Ascheberg, Meurer, & Oesterling 2011; Atwal & Williams, 2008; Yeoman & McMahon-Beattie, 2006).

Es zeigt sich, dass trotz der grundsätzlichen Eignung von mindestens einem Drittel der Reiter als Zielgruppe von Luxusmarken auch hier eine Anpassung des Marketings an das neue vorherrschende Luxusverständnis und die korrespondierenden Konsummotive erfolgen muss.

Die Anpassung der Messskalen für Luxusaffinität an die Forschungsumgebung des Reitsports erweist sich insgesamt als sinnvoll. Die Mehrfachvergleiche der Post-hoc-Analyse zeigen, dass Faktor 4 „Exklusiver Reitsport ist Luxus" bei der Abgrenzung der Segmente eine wesentliche Rolle spielt (Tab. 6). Durch die Inklusion der reitsportbezogenen Statements ist eine Annäherung an alternative Luxuskonzepte (insbesondere *luxury experience*) möglich. Es kann aufgezeigt werden, dass Luxusaffinität vor dem Hintergrund des modernen Luxuskonsums über eine rein materiell bezogene Luxusdefinition nicht länger erfasst werden kann. Eine zielgruppenspezifische Abstimmung der Messinstrumente ist auch für zukünftige Studien zu empfehlen.

Auf die Forschungsfrage lässt sich abschließend antworten, dass die These, Reiter seien im Allgemeinen eine geeignete Zielgruppe für klassisches, auf materiellen Werten ausgerichtetes Luxusmarketing, so generell nicht gestützt werden kann. Ein Wandel hin zu einem eher im-materiell ausgelegtem Luxusverständnis hat sich mindestens segmentweise auch in Hinblick auf die Reitsportler vollzogen. Eine segmentspezifische Ausrichtung des Luxus- und Reitsportmarketings ist vor diesem Hintergrund empfehlenswert.

Limitationen

Eine Begrenzung der vorliegenden Studie ist, dass bestimmte Konsumenten in der Zielgruppe der Reitsportler durch die Wahl der Umfragekanäle ausgeschlossen wurden. Aufgrund der Online-Befragung und der Hinzuziehung von zwei ausgewählten Reitsportveranstaltungen, welche zudem beide in Nordrhein-Westfalen stattfanden, ergeben sich Verzerrungen hinsichtlich der Repräsentativität.

Letztlich ist anzunehmen, dass das Phänomen der sozialen Erwünschtheit die Antworten der Probanden beeinflusst hat. Ein bis in die heutige Zeit bestehendes ambivalentes Verhältnis zum Luxusbegriff innerhalb der Gesellschaft kann dazu führen, dass Fragen zum persönlichen Luxuskonsum ablehnend beantwortet werden (Allérès, 1993; Dubois & Laurent, 1994).

Der weiterführenden Forschung sei empfohlen, vor dem Hintergrund der kulturellen Beeinflussung von Konsumverhalten (Hornig, Fischer, & Schollmeyer, 2013) Vergleiche zwischen der Luxusaffinität und den Motiven von Reitern aus verschiedenen Nationen und Kulturkreisen anzustellen. Aufgrund der florierenden Luxusmärkte und dem aufkommenden Interesse am Reitsport bei einkommensstarken Bevölkerungsgruppen in China, Russland und Arabien (Eckjans & Hartmann, 2013) ist insbesondere die vergleichende Analyse in Bezug auf diese Regionen eine wünschenswerte Ergänzung zu der vorliegenden Studie.

Auch würde eine Erweiterung der Datenbasis hinsichtlich ökonomischer Variablen Regressionsanalysen ermöglichen, mithilfe derer sich die Effekte der extrahierten Faktoren auf Zahlungsbereitschaften berechnen ließen.

Literatur

Adjouri, N., & Stastny, P. (2006). Sport-Branding im Bereich Markenunternehmen. In N. Adjouri, & P. Stastny (Hrsg.), *Sport-Branding: Mit Sport-Sponsoring zum Markenerfolg* (S. 222-232). Springer Fachmedien Wiesbaden.

Albrecht, C.-M., Backhaus, C., Gurzki, H., & Woisetschlaeger, D. M. (2013): Value Creation for Luxury Brands through Brand Extensions: An Investigation of Forward and Reciprocal Effects. *Marketing ZFP – Journal of Research and Management, 35*(2), 91-103.

Allérès, *D.* (1993). L´univers du luxe. *Regards sur l´actualité, 187,* 3-26.

Ammon, G. G. (1828). Ueber die Eigenschaften des Soldaten-Pferdes und die Mittel, die Zucht desselben zu befördern. Mittler.

Ascheberg, C., Meurer, J., & Österling, A. (2012). The Luxury Universe–Angebots-und Kundensegmentierung globaler Luxusmärkte als Basis für erfolgreiche Positionierungsstrategien. In C. Burmann, V. König, & J. Meurer (Hrsg.), *Identitätsbasierte Luxusmarkenführung* (S. 85-101). Springer Fachmedien Wiesbaden.

Atwal, G., & Williams, A. (2009). Luxury brand marketing–the experience is everything!. *Journal of Brand Management, 16*(5), 338-346.

Batinic, B., & Bosnjak, M. (2000). 11 Fragebogenuntersuchungen im Internet, Internet für Psychologen, Hogrefe Verlag Göttingen, S. 287-318.

Buchner, J. (1990). Von Pferden, Hühnern und Läusen. In I. Behnken (Hrsg.), *Stadtgesellschaft und Kindheit im Prozeß der Zivilisation* (S. 219-242). Opladen, Leske & Budrich.

Chevalier, M., & Lu, P. X. (2010). *Luxury China: Market opportunities and potential.* John Wiley & Sons.

D. R. Vereinigung (2001). IPSOS-Marktanalyse der FN zum Pferdesport. Pferdesportler in Deutschland. FN-PRESS, Warendorf URL: http://www. wpsv. de/ipsos. htm. Zugriff am 15. Juni 2006.

D´arpizio, C. (2014). *Luxury Goods Worldwide Market Study Spring 2014.* Bain and Company (Ed.), Boston.

Darlison, E. (2000): Geschlechterrolle und Sport. *Der Orthopäde, 29*(11), 957-968.

Dubois, B., & Laurent, G. (1994). Attitudes Towards the Concept of Luxury: an Exploratory Analysis. Pacific Advances in Consumer Research, *1*(2), 273-278.

Dubois, B., & Laurent, G. (1995). Luxury possessions and practices: an empirical scale. *European Advances in Consumer Research, 2,* 69-77.

Dubois, B., & Paternault, C. (1995). Observations: Understanding the world of international luxury brands: The "dream formula". *Journal of Advertising Research, 35*(4), 69-76.

Dubois, B., Czellar, S., & Laurent, G. (2005). Consumer Segments Based on Attitudes Toward Luxury: Empirical Evidence from Twenty Countries. *Marketing Letters, 16*(2), 115-128.

Eckjans, S., & Hartmann, L. (2013). Neue Märkte im Pferdesektor–China. *Goettinger Pferdetage '13: Zucht, Haltung und Ernährung von Sportpferden. Tagungsband Goettinger Pferdetage 2013.* 33-35.

Erickson, G. M., & Johansson, J. K. (1985). The role of price in multi-attribute product evaluations. *Journal of Consumer Research, 12*(4), 195-199.

Fassnacht, M., Kluge, P. N., & Mohr, H. (2013). Pricing Luxury Brands: Specificities, Conceptualization and Performance Impact. *Marketing ZFP – Journal of Research and Mangement, 35*(2), 104-117.

Flynn, B. B., Schroeder, R. G., & Sakakibara, S. (1994). A framework for quality management research and an associated measurement instrument. *Journal of Operations Management, 11*(4), 339–366.

Fournier, S. (1998). Consumers and their brands: developing relationship theory in consumer research. *Journal of consumer research, 24*(4), 343-353.

Freyer, W. (2001). Sport und Tourismus: Megamärkte in der wissenschaftlichen Diskussion. In G. Trosien, & M. Dinkel (Hrsg.), *Sport-Tourismus als Wirtschaftsfaktor: Produkte-Branchen-Vernetzung* (S. 32-65). Butzbach-Griedel.

Freyer, W. (2003). *Sport-Marketing: Handbuch für marktorientiertes Management im Sport,* 3. Aufl., Dresden.

Gamrat, F. A., & Sauer, R. D. (2000). The utility of Sport and Returns to Ownership: Evidence from the Thoroughbred market. *Journal of Sports Economics, 1*(3), 219-235.

Gille, C., Hoischen-Taubner, S., & Spiller, A. (2011). Neue Reitsportmotive jenseits des klassischen Turniersports. *Sportwissenschaften, 41*(1), 34-43.

Glare, P. G. (1982). *Oxford latin dictionary.* Clarendon Press. Oxford University Press.

Häggblom, M., Rantamäki-Lahtinen, L., & Vihinen, H. (2012). Equine sector comparison between the Netherlands, Sweden and Finland. URL: http://files.kotisivukone.com/agropolis.auttaa.fi/Uudistukset_6.2.2012/EquineLife__A_performance_model_for_an_ecologically_and_ethically_sustainable_equine_sports/equine_sector_comparison_between_the_netherlands_sweden_and_finland.pdf. Zugriff am 06. Mai 2013.

Hagtvedt, H., & Patrick, V. M. (2009). The broad embrace of luxury: Hedonic potential as a driver of brand extendibility. *Journal of Consumer Psychology, 19*(4), 608-618.

Hair, J. F., Anderson, R. E., Tatham, R. L., & Black, W. C. (1998). *Multivariate Data Analysis*, 5. Edition, New Jersey.

Han, Y. J., Nunes, J. C., & Drèze, X. (2010). Signaling status with luxury goods: the role of brand prominence. *Journal of Marketing, 74*(4), 15-30.

Hoogenraad, B., & Vering, S. (2007). Möglichkeiten des Sponsorings im Reitsport— JPMorgan Asset Management und die Nachwuchsförderung. In D. Ahlert, D. Woisetschläger, & V. Vogel (Hrsg.), *Exzellentes Sponsoring* (S. 303-322). DUV.

Hornig, T., Fischer, M., & Schollmeyer, T. (2013). The Role of Culture for Pricing Luxury Fashion Brand. *Marketing ZFP – Journal of Research and Management, 35*(2), 118-130.

Hudders, L., & Pandelaere, M. (2012). The silver lining of materialism: The impact of luxury consumption on subjective well-being. *Journal of Happiness Studies, 13*(3), 411-437.

Ikinger, C., Münch, C., Wiegand, K., & Spiller, A. (2013). Reiterleben, Reiterwelten: Zielgruppen zwischen Reitweisen, Motiven und der Liebe zum Pferd. Georg-August-Universität Göttingen, HorseFuturePanel UG, Dietz und Consorten (Hrsg.). URL: www.uni-goettingen.de/de/.../2013-04%20reitsportstudie_screen.pdf. Zugriff am 05. Mai 2013.

Institut für Demoskopie Allensbach (2013). AWA Allensbacher Markt- und Werbeträgeranalyse 2013, Allensbach.

IPSOS (2003). *Faszination Zukunft. Neue Perspektiven im Pferdesport. Die FN Marktanalyse kompakt und kommentiert.* Warendorf: FN Verlag.

Kapferer, J. N., & Bastien, V. (2009). The specificity of luxury management: Turning marketing upside down. *Journal of Brand Management, 16*(5), 311-322.

Kewes, T. (2012): Die neue Lust auf Luxus. *Handelsblatt*, 10.03.2012. URL: http://www.handels-blatt.com/unternehmen/handel-dienstleiter/luxusmarkt-waechst-die-neue-lust-auf-luxus/6308428.html. Zugriff am 13. Mai 2013.

Kisabaka, L. (2001). *Marketing für Luxusprodukte* (Vol. 32). Dissertation, Fördergesellschaft Produkt-Marketing, Cologne.

Krüger, M. (1988). Was ist alternativ am alternativen Sport?. *Sportwissenschaft, 18*(2), 137-159.

Mason, R. (1993). Cross-cultural influences on the demand for status goods. *European Advances in Consumer Research, 1*, 46-51.

Meurer, J. (2012): Ebony or Ivory – wie glänzend ist die Zukunft des Luxus in Deutschland? Kritische Reflexionen zum Luxusmarkenmanagement. In C. Burmann, V. König, & J. Meurer (Hrsg.), *Identitätsbasierte Luxusmarkenführung* (S. 321-336). Springer Fachmedien Wiesbaden.

Mohr, E. (2014). *Ökonomie mit Geschmack: die postmoderne Macht des Konsums.* Murmann Verlag DE.

Mostovicz, E. I. (2010). Satisfying the Needless Need: The Mechanism of Luxury and its Implication for Brands. *Università della Svizzera italiana (USI), Lugano, Switzerland.*

Müller-Stewens, G. (2013): Das Geschäft mit Luxusgütern. Universität St. Gallen, St. Gallen, Schweiz.

Norden, G./Polzer, N. (1995). Fernöstlicher Sport und Abendländische Kultur. In J. Winkler, & K. Weis (Hrsg.), *Soziologie des Sports* (S. 187-200). VS Verlag für Sozialwissenschaften.

Pflanz, C. (2004): Faszination Luxus. In B. Stüwe (Hrsg.), *Faszination: Marketing im Wechselbad der Gefühle* (S. 90-97). Gabler Verlag | Springer Fachmedien Wiesbaden GmbH, Wiesbaden.

Serraf, G. (1991). Le produit de luxe: somptuaire ou ostentatoire?. *Revue française du marketing,* (132), 7-16.

Statistisches Bundesamt (2013). Bildungsstand. URL: https://www.destatis.de/DE/ZahlenFakten/GesellschaftStaat/BildungForschungKultur/Bildungsstand/Tabellen/Bildungsabschluss.html. Zugriff am 27. Juli 2013.

Stegemann, N. (2006). Unique Brand Extensions Challenges for Luxury Brands. *Journal of Business Economics and Research, 4*(10), 57-68.

Strebinger, A. (2001). Die Markenpersönlichkeit und das Ich des Konsumenten: Von der Rolle des Selbst in der Markenwahl. *transfer – Werbeforschung & Praxis, 46*(2), 19-24.

Sudeck, G., & Schmid, J. (2012). Sportaktivität und soziales Wohlbefinden. In R. Fuchs, & W. Schlicht (Hrsg.), *Seelische Gesundheit und sportliche Aktivität* (S. 56-77). Hogrefe Verlag, Göttingen.

Tsai, S. P. (2005). Impact of personal orientation on luxury-brand purchase value. *International Journal of Market Research, 47*(4), 429-454.

Veblen, T. (1899). *The Theory of the Leisure Class.* New York: Macmillan.

Vigneron, F., & Johnson, L. W. (1999). A review and a conceptual framework of prestige-seeking consumer behavior. *Academy of Marketing Science Review, 1*(1), 1-15.

Vigneron, F., & Johnson, L. W. (2004). Measuring perceptions of brand luxury. *Journal of Brand Management, 11*(6), 484-506.

Wheatley, J. J., & Chiu, J. S. (1977). The effects of price, store image, and product and respondent characteristics on perceptions of quality. *Journal of Marketing Research, 14*(2), 181-186.

Wiedmann, K. P., Hennigs, N., & Siebels, A. (2007). Measuring consumers' luxury value perception: a cross-cultural framework. *Academy of Marketing Science Review, 7*(7), 333-361.

Wiedmann, K. P., Hennigs, N., & Siebels, A. (2009). Value-based segmentation of luxury consumption behavior. *Psychology & Marketing, 26*(7), 625-651.

Wieseke, J., Alavi, S., Habel, J., & Dörfer, S. (2013). Erfolgsstrategien im persönlichen Verkauf von Luxusmarken. *Marketing ZFP – Journal of Research and Management, 35*(2), 131-143.

Wong, N. Y., & Ahuvia, A. C. (1998). Personal Taste and Family Face: Luxury Consumption in Confucian and Western Societies. *Psychology and Marketing, 15*(5), 423-441.

Yeoman, I. (2011). The changing behaviours of luxury consumption. *Journal of Revenue & Pricing Management, 10*(1), 47-50.

Yeoman, I., & McMahon-Beattie, U. (2006). Luxury markets and premium pricing. *Journal of Revenue and Pricing Management, 4*(4), 319-328.

I.2 Combining One-to-One-Marketing and High-End Luxury: Theory-building from Customized Luxury Saddles for Chinese Horse Riders

Authors: **Laura Hartmann, Achim Spiller**

Georg-August-University of Goettingen

This article is published in a short version in *Proceedings of the 2014 Global Marketing Conference at Singapore. Bridging Asia and the World: Globalization of Marketing & Management Theory and Practice.* ISSN number: 1976-8699.

Abstract

Recently, luxury markets have been flourishing in both East Asian and Western cultures, and some are predicting a further upward trend. At the same time, there is a shift of motives for luxury consumption from prestige-seeking and conspicuousness toward hedonism, experience and self-expression. Accordingly, marketing strategies for personalization and mass-customization are emerging in order to serve consumer demand for individualized products. What remains uncertain is how individualization can be realized in luxury markets in its strictest meaning, i.e. markets in which the highest price levels offset low quantities. We find hypothetical answers by examining a theory-building single-case study. Based on an analysis of the Chinese market for customized and handmade horse-riding saddles and considerations of cultural peculiarities and the development in consumer behavior, we develop a framework that links one-to-one marketing with high-end luxury.

Keywords

Luxury, Case Study, China, One-to-One Marketing

Introduction

In less than a decade, China has gone from an almost insignificant consumer society to a key market for global luxury brands (Chevalier & Lu, 2009; Garner, 2005). The pace of transformation is stunning, and this also applies to the equestrian market. Demand for luxury experience is expected to occur almost simultaneously with the demand for luxury goods (Atwal & Williams, 2008; Dubois & Laurent, 1995, Wiedmann, Hennigs, & Siebels, 2009). Thus, expensive sports like horse riding, golfing and skiing have seen a huge surge (Eckjans & Hartmann, 2013). The import volume of sport horses to the Chinese market has increased by more than 1400% within the last six years. The equestrian market has also developed quickly in terms of quality. Equestrian education and the professionalism of horse-riding events are improving continuously. This results in an increased demand for high quality equipment and services. Among other things, there is a rising demand for customized riding saddles among ambitious Chinese riders.

Luxury consumers in China and horse riders in general are known to have some common characteristics. First, both groups are motivated by hedonism and self-fulfillment as well as success and acceptance (Gille, Hoischen-Taubner, & Spiller, 2011; Lu, 2008). Second, their socio-demographic profiles have major similarities. Generally speaking, Chinese luxury consumers and horse riders are highly educated, high-earning young people, and both communities have an almost balanced gender distribution.[6]

These findings lead us to the assumption that customized saddles can be used as an example of practicing one-to-one-marketing (Peppers, Rogers, & Dorf, 1999; Peppers & Rogers, 1997) in the context of luxury industries. Individualization strategies have been observed as an increasingly effective instrument within Chinese culture. This serves as a base for the definition of a new marketing concept, standing out from classical strategies of luxury marketing. In order to better understand the relationship between one-to-one-marketing and luxury markets, we examine a case study in combination with market research in Beijing and Shanghai. We surveyed a French manufacturer of luxury products and customized handcrafted leather saddles and focused on the firm's operations in the Chinese market, its target group and associated cultural factors. The study is intended to motivate a new marketing concept at the interface between luxury goods marketing and one-to-one-marketing. The knowledge generated in the

---

[6] For socio-demographic information on Chinese luxury consumers see Roland Berger Strategy Consultants (2012).

course of the analysis is, therefore, used to derive implications for the luxury industry in general, within the Chinese setting and internationally.

We address whether one-to-one marketing, which has emerged as a new marketing strategy and research field in marketing sciences with growth in the use of the Internet by companies for selling and customer communication (Pitta, 1998), is a suitable concept for marketing luxury goods in China. We further ask whether it fulfills the Chinese luxury customer's needs, and how far the Chinese case may be generalized to luxury marketing in other cultures. In order to investigate these issues and to systemize our concerns of research, we use a case study to address the following four research questions.

(1) How can the idea of one-to-one marketing be realized in luxury goods?

(2) Is the case of high-quality, customized riding saddles sold in China an appropriate example where one-to-one marketing and marketing for luxury goods are linked?

(3) What are the implications of this case for other luxury industries?

(4) Can we transfer conclusions about Chinese luxury customers to target groups in other cultures?

Darke, Shanks, and Broadbent (1998) and Eisenhardt (1989), who propose guidelines and a conceptual framework for building theories from case study research, inspire our approach. Accordingly, we interpret the case study results led by the research questions stated above and shape the hypotheses. Although both one-to-one marketing and high-end luxury are well-defined marketing strategies, they rarely have been researched in combination. We therefore generalize our findings to a theory that has to be tested and confirmed in the future, e.g. by studying multiple cases.

Research methods

The triangulation of data sources applied in case study research allows developed theories to be tested. The common use of multiple data collection methods is known to generate positive synergistic effects. Quantitative research can reveal relationships within data while qualitative methods are used to explain them and build theories (Eisenhardt, 1989). Our case study is based on four different channels of data accumulation. First, a standardized questionnaire targeting Chinese participants in equestrian activities was quantitatively evaluated. Second, face-to-face interviews with Chinese equestrians provided us with quantitative information. Third, the company in the case was analyzed both quantitatively and qualitatively, and fourth, we built upon our knowledge of broad experiences in the Chinese equestrian scene, which we acquired by observing the market for one year.

Quantitative statistics are purely descriptive and used to support our assumption that substantial links between luxury marketing and one-to-one marketing can be found in our case. Observations, interviews and data from archives help to explain the underlying concepts.

Our field research was completed during a journey to Beijing and Shanghai in June 2013. We visited eight riding stables in Beijing, five in Shanghai and two national competitions. Meanwhile, we interviewed stable managers, riders and riders' associates. Sixty-seven randomly chosen people we met with face-to-face at the stables or competitions filled out the questionnaire. Quotas were not specified. It contained 34 questions and was subdivided into four parts. First, people were asked about their involvement in horse sports and associated reasons; second, about their attitude and habits with regard to customized saddles and sale processes; third, about their general attitude toward luxury; and fourth, about their socio-economic data. In the first and second part, information on willingness-to-pay for horse sports in general and equipment in particular were included. The questions concerning saddles and sales services were developed in collaboration with a French manufacturer of both luxury goods and high-end customized saddles. We formulated 14 statements for evaluation by means of a five-point Likert scale (from full agreement to no agreement at all) and combined it with closed and half-closed questions with multiple answering options.

The firm was chosen in accordance with Flyvbjerg (2006) and Pettigrew (1988), who proposed to analyze extreme cases where the research object of interest could be observed paradigmatically. According to Flyvbjerg (2006), extreme cases are often more significant than representative cases or random samples if the research aims to investigate the causes behind a given problem at depth. In this sense, the phenomena of luxury and demand for individual products are presented in their deepest meaning in this firm's products. They represent luxury at high price ranges and a narrow interpretation of individualization. The case may thus provide an understanding for the underlying causal links between the demand for customization and the motives for luxury consumption.

The techniques to survey the respondent's affinity toward luxury in general were inspired by a cross-border study of attitudes toward the concept of luxury by means of an attitudinal scale

(Dubois, Czellar, & Laurent, 2005) and modified according to our research environment. We thus included six attitudinal statements for evaluation using a five-point Likert scale.

The data from the questionnaire is analyzed descriptively. The aim was to provide a first, quantitative overview of the attitudinal and socio-demographic characteristics of Chinese horse riders in classical riding since this field of research has been completely neglected in international literature. The descriptive information is applied to the knowledge and our interpretations derived from the case. We interviewed employees (non-standardized) and analyzed the company's recent financial statement as well as reports on the firm published in popular or scientific media (Chevalier & Lu, 2010; Focus, 2009; Lu, 2008). Thereby, we concentrated on the firm's operations in customized riding saddles since these products fulfill our research conditions—the combination of luxury and customization. The earliest steps of the firm's entry into the Chinese equestrian saddle market were observed.

Description of the case company and its markets

Customized saddles have been popular for some time among ambitious European riders, at least in the classical disciplines like dressage or jumping, in polo sports and endurance. The level of customization may differ substantially depending on the different manufacturing stages in which adjustments are done. It varies from ready-made saddles where just the cushion is adjusted to fit a horse's back to the extreme case of completely individually tailored saddles. The intensity of labor and acquired knowledge depends on those customization levels. While, for example, adjusting the cushion can be mostly done by experienced sales persons in a specialized shop, full customization is time-consuming, sometimes taking up to half a year and requiring highly-skilled saddlers. Full customization means that horse and rider are measured, the sitting position of the rider, as well as known problems and potential for optimization, have to be assessed and information about the rider's preferences with regard to comfort level, leather quality, weight and design features have to be gathered. Based on those details, a saddle is handcrafted. Complete machine fabrication and production of higher quantities is not possible. Due to these requirements, the supply side of the market is very small while the demand is comparatively large. As far as we know, no more than five manufacturers of handcrafted saddles operate internationally. To get a basic impression of the number of potential customers, we used the database from the international umbrella organization for equestrian sports, the FEI. They state that around 74,000 riders are registered to compete at international horse shows, which means that they are professionals or at least semi-professionals (FEI Database, 2013). These riders correspond to the target group for customized saddles. The lower the degree of customization, the more firms supply the market. Manufacturers of ready-made saddles compete on an international level against many rivals from all over the world.

This case study examines one of the few well-known international saddle-makers whose products are completely handcrafted and consistently made from high-valued materials, including the latest innovations. The firm was founded in 1837 in Paris and has been regarded as a producer of handcrafted saddles since 1867. Over time, they have expanded their range of sectors and now supply general leather goods, ready-to-wear clothes, silks and textiles, jewelry, furniture, furnishing fabrics, wallpaper, tableware, perfumes and watches. Furthermore, the company was chosen to provide the interior decoration of a private jet and a 25m yacht. Despite this diversification, they still treat equestrian products and leather articles as their traditional business. All the products that they market are among the most expensive in each respective sector. Their saddles in particular are the most expensive on the market.

The firm produces eight basic types of saddles, which differ according to discipline, technical details and design. For every saddle, the customer may choose between color, size, type of leather and some technical features. Even though the saddles are basically ready-made, some adjustments can be made to the cushion or to the saddle tree in order to fit the saddle to the horse's back. For these services, the company's saddler visits the stable where the horse is accommodated. The prices for these saddles range from 4,050 Euro to 6,700 Euro. At the other end of the price spectrum, the company offers completely customized saddles as described above. For this, the saddler again comes to the customer's stable, takes measurements and gathers all necessary information. The prices depend on the customer's choices regarding leather quality, technical features and design, but the saddles begin at 7,000 Euro. Overall, the company's revenue from saddlery and leather goods amounted to 1.6 billion Euros in 2012.

The company supplies 25 countries all over the world with 323 exclusive stores, but the production of saddles is completely located in Paris. Saddles and other equestrian products are sold in every shop along with the other products belonging to the company's brand. Thus, the various products of the brand are marketed abreast. Nevertheless, small differences appear in the distribution channels. On the company's website, only photographs of the saddles, technical information, corresponding options and contact information of the responsible saddler are published. While other equestrian products from that brand, like bridles and blankets, can be bought on the website, this is not possible for saddles. The same applies to jewelry.

Our focus on China is justified by current market developments, which are directing the attention of the European equestrian industry toward Chinese riders. Accompanied by a general trend seen in recent years within China for luxury goods and luxury lifestyles, especially with regard to prestigious European brands (Atsmon, Ducarne, Magni, & Lu, 2012; Chevalier & Lu, 2010; Lu, 2008), horse sports are becoming more and more popular (Eckjans & Hartmann, 2013; Zhou & Hui, 2003). European brands of horse-riding equipment and breeding associations are realizing their market potential and integrating concepts for market entry into their short- or mid-term strategies. This also applies to our case study. Customized saddles from the French manufacturer should be placed into the Chinese equestrian market, especially in Beijing and Shanghai. They expect their strategy to develop successfully, based on the assumption that there are major overlaps between the target groups and consumer motivations in both the horse sports and luxury industries.

Large differences in income and the emergence of a very rich, highly educated and western-orientated social class in China have been repeatedly described in the literature and a section of wealthy Chinese consumers has shown to have strong preferences for luxury goods (Atsmon, Ducarne, Magni, & Lu, 2012; Garner, 2005; Lu, 2008; Lu & Pras, 2011). According to a consumer report published by CLSA Asia Pacific Markets (2011), the luxury industry had a turnover of 9.2 billion Euro in 2011. Its compound annual growth rate (CAGR) is estimated to be 25% over the next five years, compared to a CAGR of general consumption of 11% in the same time period. In 2011, luxury sales in greater China amounted to 10% of the global market. The 2020 market share is forecasted to be 44%. If this happens, it would be the largest market for luxury goods worldwide.

The company in this case study has already anticipated the upward trend in the Chinese luxury industry. With 20 stores opened by 2012, the brand's products like ready-to-wear clothes, handbags, luggage and jewelry are already selling successfully. In Asia (Japan excluded), the turnover was 1,100 million Euros. This means 32% of the company's overall revenues in 2012 and hence, Asia (Japan excluded) is the company's key market. The growth rate compared to 2011 amounted to 25% at constant exchange rates. Leather goods and saddlery, which are combined in the company's financial statement, generated revenues of 1,597 billion Euros worldwide in 2012 (with a growth rate compared to 2011 of 12% at constant exchange rates). It is the brand's most successful division in terms of revenue.

By contrast, the sales of equestrian equipment in China remain comparatively low due to strategic deficits in the firm's marketing. Chinese riders do not seem to be aware of the brand as a producer of high-quality saddles and tack, even though there is significant potential in the market. We find evidence for the latter in our field research. Respective results are discussed in the following section.

Findings from field research and discussion

Our questionnaire- and interview-based market survey revealed that Chinese riders belong to high or the highest income groups. Their willingness to pay for equestrian products is therefore high. In the survey's random sample, more than one quarter spend on average more than 1,500 Yuan (approximately 178 Euro)[7] per month on riding equipment. For horse sports in general, a share of 19 % stated to spend more than 12,000 Yuan (approximately 1,422 Euro) per month. This is high compared to Germany. For example, D. R. Vereinigung (2001) finds that German riders spend on average 203 Euro per month on horse sports. According to a more recent but not yet published survey of German equestrians that we conducted in 2013, only 10% of riders spend more than 1,500 Euro per month on horse sports.

In the questionnaire, we asked why people are interested in doing horse sports. The aim was to find out how far they associate horse sports with traditional luxury motives. It becomes apparent that Chinese equestrians rather do not look for an expression of prestige nor are they focused on establishing contacts or meeting like-minded people when joining in their sport (Figure 1). Instead, they seem to be mainly motivated by love for animals (68% of all participants) and playing healthy sports (49% of all participants). These results imply the following: Horse riding is not associated with socially-oriented luxury consumption. On the contrary, people seem to distance themselves deliberately from the idea to only connect horse sports to conspicuousness. Social status and collectivism are not the major motives to join the equestrian community. Rather, riders intend to express hedonistic, emotional and individualistic values. Doing something good for one's constitutions, enjoying life and combining this with the passion for horses are found to be strong motives for equestrian activities. However, this is similar to findings from research on the motives of German horse riders (Gille, Hoischen-Taubner, & Spiller, 2011; D. R. Vereinigung, 2001).

---

[7] Exchange rates are derived from the European Central Bank.

(http://www.ecb.europa.eu/stats/exchange/eurofxref/html/index.en.html, last accessed 24 October 2013).

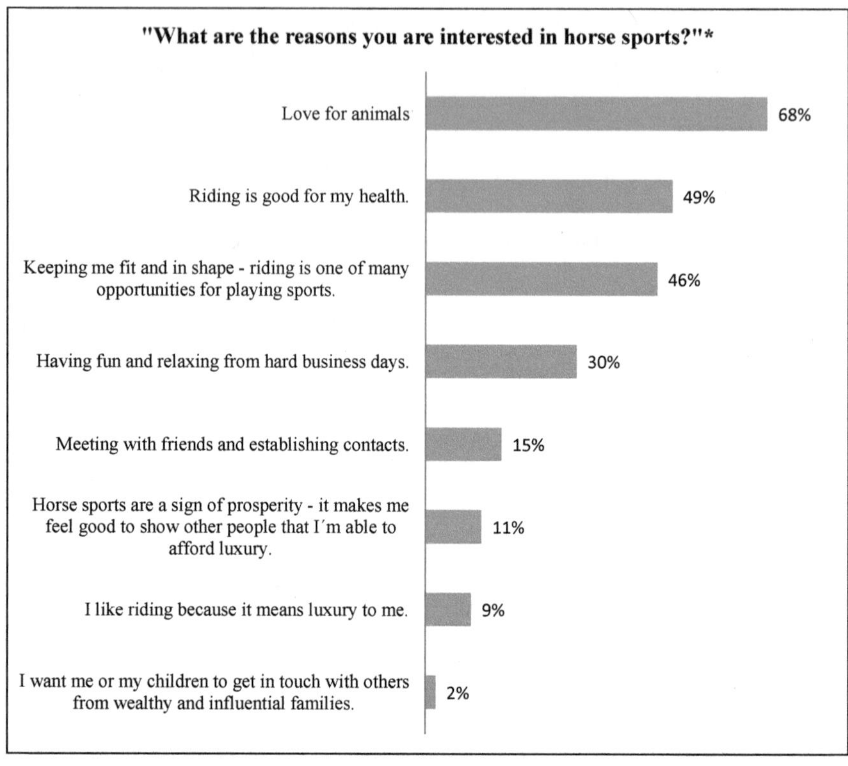

Figure 1. Answers given to a question concerning reasons for doing horse sports

(*multiple answers were possible; rounded values)

Our surveys showed that the attitude of Chinese riders toward the importance of saddles and their willingness to pay for them is advantageous for providers of customized saddles, especially from Europe. Simultaneously, customer demands regarding sales services and buyer-manufacturer relationships was shown to represent a challenge for practical marketing. We asked people to name the most important pieces of equipment in their opinion. Further, participants evaluated for which pieces of equipment they accept paying a higher price for better quality. From eight different answer options given, in both cases participants mostly chose saddles (Figure 2). For the question in which country do the best saddle makers originate from, Germany, France and Italy were each named by more than 50% of the sample (Figure 3). In other words, the participants mostly trust the quality of saddles from those nations that are strongly associated with handcrafted, high-quality luxury products. Among the first four most valuable luxury brands in 2012 are two German (rank 1 and 2), one French (rank 3) and one Italian (rank 4). Furthermore, LVMH, the leading producer of material luxury goods in terms of sales, is originated in France. All three countries are known for their successful luxury industries (Gabel, 2013). The appeal of European luxury brands to Chinese luxury consumers is found by Atsmon, Ducarne, Magni, and Wu (2012). Especially tradition, origin and quality are highly valued and perceived to be typical for European luxury brands (Chevalier & Lu, 2009; KPMG, 2008; Lu, 2008).

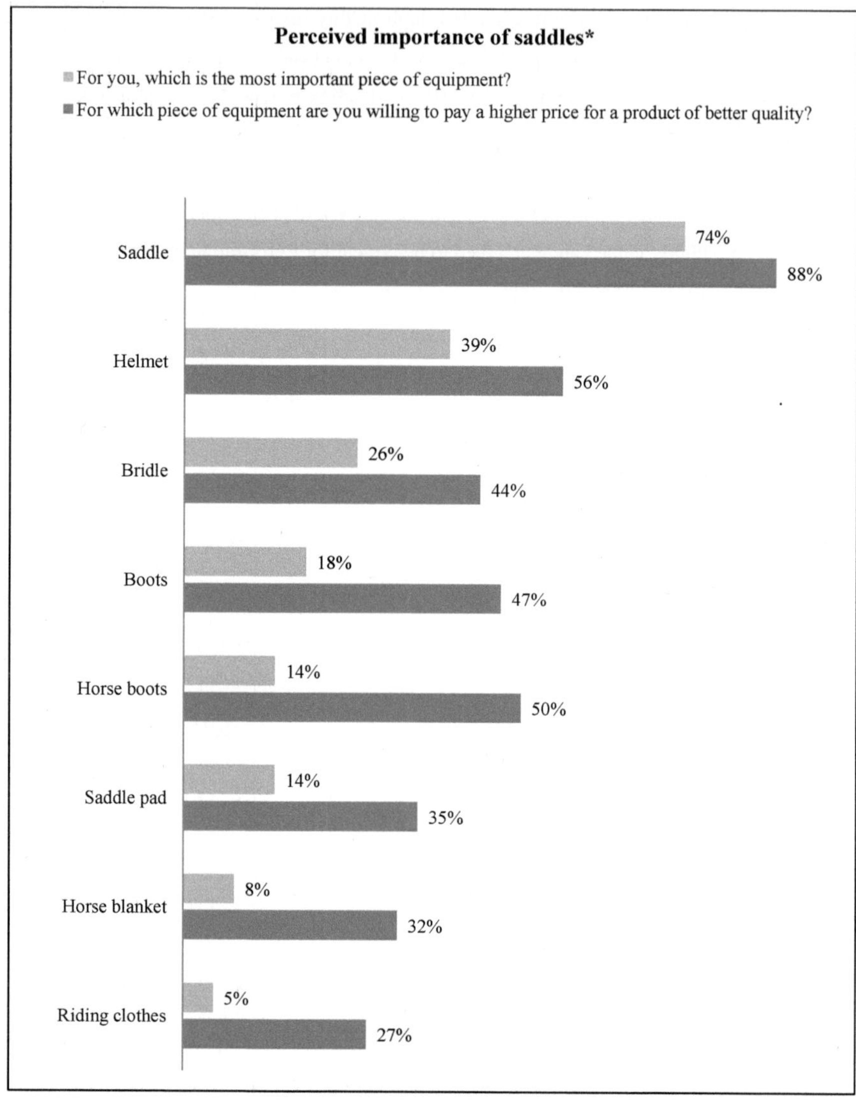

Figure 2. Answers given to attitudinal questions about horse riding equipment

(*multiple answers were possible; rounded values)

Further, the questionnaire contained 14 statements about saddles and associated sale processes, which were evaluated by means of a five-point Likert scale (from full agreement to no agreement at all). For example, more than 50% fully agreed with the statements "For me, trust in the manufacturing firm and its products is a very important aspect when buying a saddle" and "Normally, the saddle should be fitted on the horse's back before buying" and "An unsuitable saddle makes it harder for the rider to have a proper posture" (Figure 4). In summary, saddles in general are perceived to be an important piece of equipment and their quality is considered to be crucial for the health and performance of both horse and rider. Two statements were used to address the customer's demand for purchase processes.

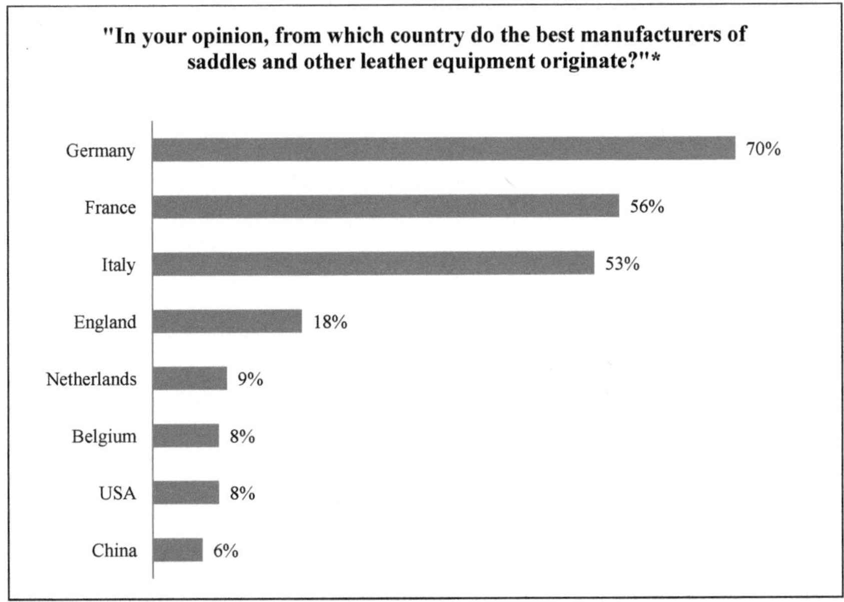

Figure 3. Answers given to the question concerning the best origin countries for saddles

(*multiple answers were possible; rounded values)

The evaluation of the statements about saddles and sale processes revealed that the majority of potential buyers of saddles are interested in having a close relationship with the manufacturing firm. Sales services and purchase support are demanded. Fifty-three percent fully agreed with "Trust in the manufacturing firm and its products is a very important aspect when buying a saddle." Furthermore, 61% fully agreed with the statement "Sales and after-sales services (professional consultancy, customer support, informative meetings, customer events, etc.) are very important to me" and 65% fully agreed that "It would be great if I could test a saddle by riding with it before I buy it." Correspondingly, participants seem to be unsatisfied with the current situation in the Chinese market for saddles. Considerably fewer people positively evaluated statements regarding the quality of specialized shops and purchase support. "Sales assistants can give good advice on the right saddle for the customer" was fully agreed with by 14% of participants. "Sales associates know all the details about different saddles" was fully agreed with by 12%, and "There are many shops in China that sell saddles. It is easy to find a good quality saddle." was fully confirmed by 11%.

In summary, our results confirm the assumption that demand and supply on the Chinese market for high quality saddles is unbalanced. The customer's wish for competency in sales services and a trust-based manufacturer-supplier relationship create a market gap that could possibly be filled by the customization of saddles. Relationship marketing and personalization of sales processes appear as key success factors for companies on the Chinese market for saddles.

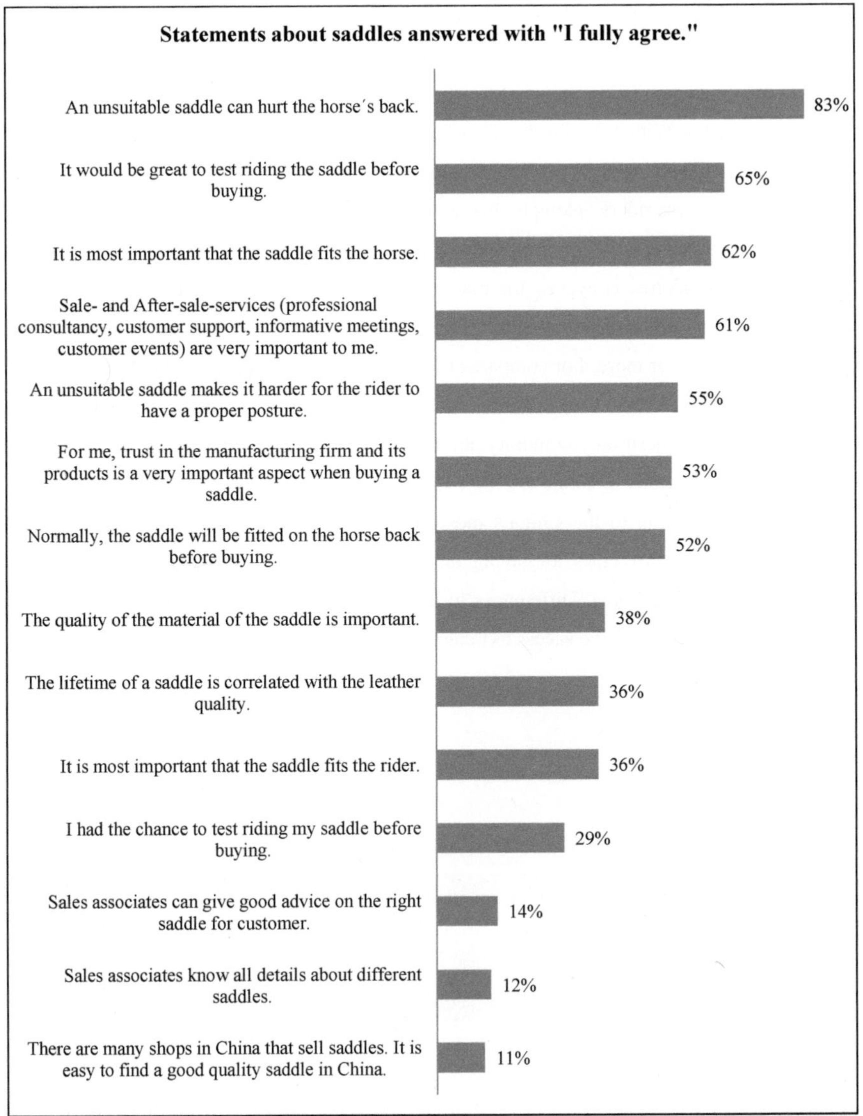

Figure 4. Evaluation of Likert-scaled statements about saddles (*rounded values)

With the help of a market survey of the luxury industry in China by Roland Berger Strategy Consultants (2012), we can compare the socio-demographic characteristics of both consumer groups (horse sports and luxury). Both industries are found to have a similar gender distribution, with only small deviations (Figure 5). While luxury consumers are slightly more often women than men (55% versus 45%), the equestrian industry is slightly male-dominated (42% versus 58%). With regard to the distribution of age within both industries, there is a concentration of people aged 20 to 39 years old. Forty-one percent of Chinese luxury customers and 60% of Chinese riders belong to this group (Figure 6). For income, estimations are less practical, because only 54% of respondents gave us information about their average annual net household income. However, the majority of participants who gave income information belong to the high-income category. Twenty-six percent stated that they earn 120,000 Yuan (14,197 Euro) or more. For comparison, the average annual net household income in China is 25,400 Yuan (3,005 Euro) (Atsmon, Magni, Li, & Liao, 2012). Based on our observations in the Chinese equestrian community and the fact that the costs of keeping horses and membership to horse clubs in China are high compared to average income levels, it can be assumed that riders belong to the China's most affluent social class. Atsmon, Magni, Li, and Liao (2012) define this class as having an annual disposable household income that exceeds $34,000 (24,632 Euro). Differences in income are large, meaning that the Chinese lower or middle class do not have access to luxury and horse sports markets (Lu, 2008).

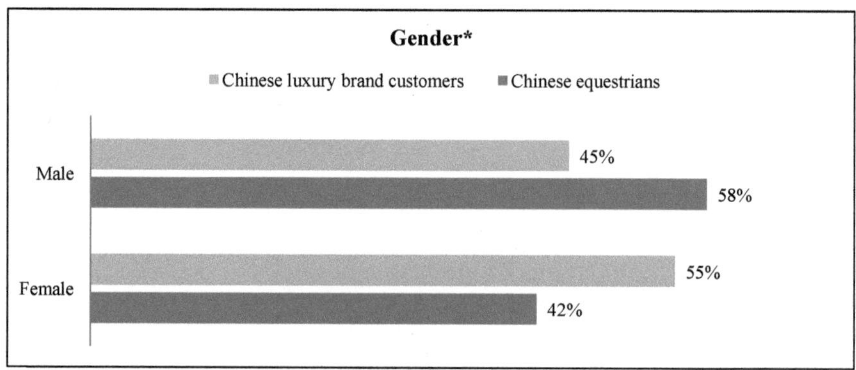

Figure 5. Gender distribution within the sample[8] (*rounded values)

---

[8]Additional sources: Roland Berger Strategy Consultants (2012).

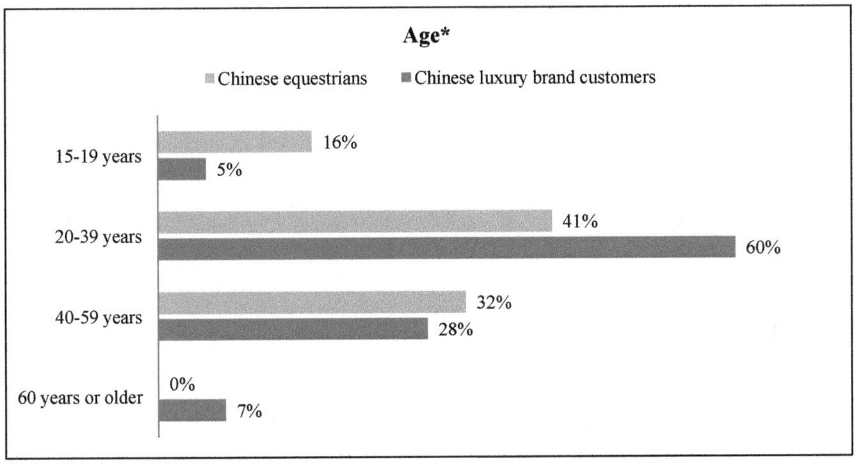

Figure 6. Age distribution within the sample[9] (*rounded values)

In the Chinese luxury market, 68% of customers belong to the country's richest 20%. Only 18% earn a medium income (vs. 30% of the total Chinese population), and 14% have a low income. In the total population, a low income is earned by 50%. Most luxury consumers are either private business owners, corporate executives, wives of wealthy men or belong to the wealthy second generation. The situation among riders is similar. Forty-six percent of all participants were company owners, executives, general managers or non-working wives (Roland Berger Strategy Consultants, 2012).

The attitudes of the surveyed Chinese riders regarding luxury correspond to their socio-demographic data. Approximately one-third evaluated the statements "Basically, I like luxury", "I inform myself regularly about the latest trends in fashion, cosmetics, furnishing and lifestyle", "Luxury makes life enjoyable", "When I spend time in shops for designer brands, I feel happy" and "I enjoy talking with friends about newly purchased goods, fashion, design and lifestyle" positively. Shares of participants who particularly did not agree with this were below 20% for each statement. Thus, a predominant affection for luxury was revealed.

At this point, social desirability should be considered in order to prevent mistakes when interpreting questionnaire results due to luxury. This phenomenon may lead participants to evaluate statements with "Partly/partly" or "No comment", or even state attitudes that are untrue.

---

[9]Additional sources: Roland Berger Strategy Consultants (2012).

Since humility and modesty are deeply rooted in the *Confucian*-influenced Chinese value system, luxury is often consumed discretely. This may prevent people from disclosing information about personal consumption of luxury, since they may find themselves in attitudinal ambivalence (Lu, 2008). Thus, we assume that social desirability weakens here the real affinity of equestrian people toward luxury goods.

Summarily, we found evidence of a positive attitude toward luxury goods within the group of our participants. This is consistent with the findings of several other studies on luxury consumption in Asian cultures (e.g. Chevalier & Lu, 2009; Lu, 2008; Lu & Pras, 2011). According to our results, Chinese equestrians consume luxury goods and have no negative association with others who own luxury products. It is also important to note that they do not consume luxury for social status or prestigious consumption motives. Instead, they are driven by hedonistic and personally-oriented motives like enjoyment and self-fulfillment. Immaterial values like time and health are highly valued and belong to the definition of the multi-dimensional construct of luxury. We concluded that Chinese equestrians demand luxury brands but they are reluctant to connect it with prestige and conspicuous consumption. Congruently, many participants fully agreed with the idea that "real luxury" is discreet and should remain unobtrusive. The particular item was fully approved by 60% and further, 28% slightly confirmed it (Figure 7). This attitude corresponds to the findings of Atsmon, Ducarne, Magni, and Wu (2012), who investigated the attitudes toward luxury of the general Chinese luxury customers. They found a new preference for understated luxury products among Chinese luxury buyers. More than half of this target group found that showing-off is in bad taste (51%) and preferred less flashy products (66%). These shares had grown rapidly; they were 37% and 50% in 2010. The authors further report that Chinese luxury consumption has transformed, and they confirm the intensifying relevance of individualistic, self-directed luxury motives.

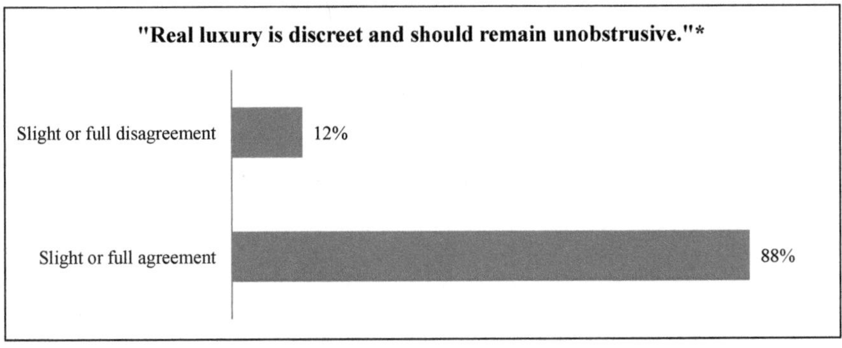

Figure 7. Likert-scaled evaluation of a statement about luxury (*rounded values)

Thus, we observe a need for marketing concepts which respond to the customer demands revealed here. Traditional luxury values like prestige and conspicuousness, which are directed to social stratification and demonstration of wealth, have obviously shifted into the background. Modern self-directed motives for luxury consumption have instead gained importance and are nowadays predominant in the target groups of Chinese equestrian people and Chinese consumers of high-end luxury products. Different literature on Chinese luxury markets confirms our findings (Atsmon, Ducarne, Magni, & Wu, 2012; Chevalier & Lu, 2010; Degen, 2009; Lu, 2011).

Defining luxury and one-to-one marketing and applying it to the case study

Luxury

Many aspects of consumer perceptions and motives related to luxury have been investigated in social sciences and economics. Despite the recognition of luxury as a demonstration of wealth, social position, conspicuous consumption and prestige (Mason, 1993; Veblen, 1899), newer studies refer to a shift from luxury as status goods toward immaterial consumption motives like hedonism, personal fulfillment and experience (Ascheberg, Meurer, & Oesterling, 2012; Atwal & Williams, 2009; Yeoman, 2011; Yeoman & McMahon-Beattie, 2010; Yeoman & McMahon-Beattie, 2006). Additionally, perfectionism, here meaning a predilection for highest quality, is identified to be part of the motivational background of luxury consumers

(Dubois & Laurent, 1994, Vigneron & Johnson, 2004; Vigneron & Johnson, 1999; Tsai 2005).

Wiedmann, Hennigs, and Siebels (2009) summarize that "[...] luxury is a subjective and multidimensional construct [...]" whose definition "[...] should follow an integrative understanding [...]" (p. 627). They name four different dimensions of luxury value: financial, functional, individual and social.

Luxury is consumed by means of luxury goods, whereas the answer to the questions how a luxury good can be exactly defined and where the borderline is between luxury and non-luxury goods depends on the viewer's perspective. Luxury goods are often described as having a high prestigious value, which in turn is mainly indicated by certain brands and their recognition in society, and high prices (Fassnacht, Kluge, & Mohr, 2013; Hellhammer, 2009; Imobersteg, 1967). Albrecht et al. (2013) and Dubois and Paternault (1995) refer to the "dream value" as a characteristic describing the consumer's strong desire to own particular items of one or another luxury brand.

Nevertheless, the definition of those goods which mean luxury for an individual person is highly subjective. For example, it can be differentiated between material and immaterial luxury goods (Pflanz, 2004). In general, the description of the construct luxury is often based on consumption motives. They are not only influenced by culture, but they are also closely linked with a change in value systems over time (Wong & Ahuvia, 1998). Correspondingly, marketing strategies should mirror the actual motives of customers for luxury consumption. Continuous research on consumer behavior is demanded in order to provide the necessary information. Only then can marketing strategies adapt to specific target groups (Ascheberg, Meurer, & Oesterling, 2012).

Further practical implications for marketing luxury goods are derived from numerous scientific studies. For example, the enhancement of a brand's hedonistic potential is recommended in order to achieve favorable brand extension (Hagtvedt & Patrick, 2009). Other authors advise segmenting luxury customers and suppliers instead of assuming homogenous target groups in order to successfully position a luxury brand (Ascheberg, Meurer, & Oesterling, 2012) or to create conditions for word-of-mouth sales because reputation is identified to be crucial for a luxury brand's success (Danziger, 2005). As we will later see, segmentation is an important tool in marketing that reaches its extreme form in one-to-one marketing (Arora et al., 2008).

Definitions of luxury from the "brand point of view" address premium consumer goods within the high price range that are produced in large quantities, like ready-to-wear-clothes, leath-

76

er articles and perfumes from designer brands. In contrast, Kapferer and Bastien (2009) claim a much more concise demarcation. Luxury products are those that are unique, handcrafted, extraordinary and with "a very strong personal and hedonistic component" (p. 4). They are produced in a "qualitative and not quantitative" manner (p. 5), whereas hedonism and the demonstration of social superiority push functionality into the background.

In this study, we refer to a similar definition as stated by the latter authors. Customized riding saddles from the manufacturer in our case are handcrafted and unique. Their quality is known to be outstanding and their prices are the highest in the market. Hence, mass-production is not possible and besides, would be useless since the customer base is small compared to consumer goods like apparel. What remains different from the definition of Kapferer and Bastien (2009) is the aspect of usability. Customization in the context of this case study is shown to be highly desirable from a functional point of view, as the physiological meaningfulness of well-fitting saddles can be seen in veterinary literature (Cocq, Weeren, & Back, 2004; Dyson, 2000). Accordingly, the motivation for buyers of customized saddles in the highest price range is more complex. Prestige and hedonism play a subordinated role. Instead, buyers seem to be aware of a saddle's dimension of functionality. The fit of the saddle for rider and the horse is highly important and can be assumed to have more influence on purchase decisions than product dimensions that demonstrate prestige or hedonic motives. This assumption is supported by the scoring of the statements in the questionnaire referred to above. The statement "It is most important that the saddle fits the horse." was fully agreed to by 62% of all participants and the statement "An unsuitable saddle can hurt the horse's back." even found 100% approval by 83% (Figure 4). A further question in the study among Chinese riders reaches a similar conclusion. The participants were asked to state what they thought are the most important product characteristics of riding equipment (saddles included). While only 3% and 22% chose the answer "That it is expensive." and "That it is beautiful.", respectively, most answered "That it fits the horse.". Sixty-six percent were mostly concerned with quality, and 63% found it important that the tack is functional (Figure 8).

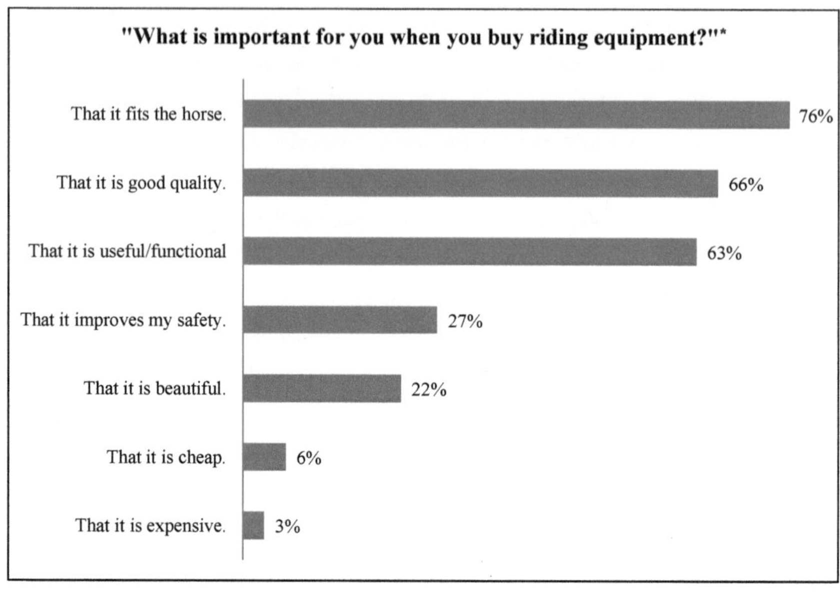

Figure 8. Answers given to a question concerning characteristics of equipment

(*multiple answers were possible; rounded values)

Since we find evidence that quality matters more in this case than prestigious or self-expressing product functions like social stratification, the definition of luxury shifts here to handcrafted, customized products whose production requires highly skilled professionals, whose components have outstanding quality and that are sold for the highest prices in the market. Functionality, especially with regard to its fit, is perceived to be the crucial product characteristic. Customers are willing to pay a high price in order to receive a quality (functionality) mark-up.

One-to-one marketing

One-to-one marketing is a concept that is based on personalization and customization strategies (Arora et al., 2008). Both refer to the adaption of one or more marketing policies to information the company has on its customers' preferences. The main difference is that in personalization strategies, firms proactively gather information about the customers' needs and decide on the most suitable marketing mixes applied, while customization means that customers act proactively as they specify by marketing dimensions like product design and price, distribution or communication policies themselves. The degree to which companies practice personalization can differ in a range between both extreme forms—mass-market marketing and one-to-one marketing. In the first, no personalization is applied because all customers are served equally, e.g. advertisements are sent to all households in a town. In the latter, every single customer is addressed personally, e.g. individual invitations and letters. Hybrid forms split the market in more or less homogenous customer groups and treat them equally. Evidence can be found in literature that customers perceive personalized marketing to be more useful than standardized marketing policies as they face lower perceived transaction costs and positive emotional feelings (Liang, Chen, & Turban, 2009). The main challenge of personalization lies in the adoption of technologies ensuring that customer data is gathered, analyzed and managed efficiently. Personalization relies heavily on information technology, which creates an information base for tailored products (Pitta, 1998).

In the case of customization, the extent to which marketing tools are specified by customers ranges from standardized products in a mass market (no customization) to completely tailored goods. The chosen extent of customization influences to what extent single steps within the production processes can be shared across products. In traditional product differentiation as a weaker form, firms use distinct product features and address more or less narrow target groups in order to differentiate themselves from other suppliers or to provide variety in their own product range. Production processes are standardized as far as possible (Smith, 1956). Complete customization, in contrast, requires individual production processes for every single product, e.g. hand-tailored dresses. An example for hybrid forms is cars. In this case, customer choices are mixed with standardized product parts (Scheer & Loos, 2001).

What matters is the cost, which becomes higher as the level of standardization in production decreases. Here, the challenge is to choose a degree of customization where its associated strategic advantage overcomes the mark-up on production cost (Arora et al., 2008). The strategic advantage consists of enhanced customer satisfaction. Examples from the apparel, sporting and technology industries show that customers prefer customized to standardized products (Berger & Pillar, 2003; Feitzinger & Lee, 1997; Pitta, 1998).

Both concepts, personalization and customization, are based on market segmentation, which is found in its extreme form in one-to-one marketing. Again, mass marketing represents the exact opposite, as it addresses broadly defined or even undetermined market segments (Arora et al., 2008). In order to serve customers individually, one-to-one marketing has the goal of transforming comprehensive information about a particular customer's demand into perfectly tuned products and services. This means the entire value chain within the manufacturing firm has to be coordinated accordingly (Peppers, Rogers, & Dorf, 1999).

How can we classify customized and handcrafted saddles into a two-dimensional space formed by the axes of personalization and customization? In terms of customization, the saddle can range from being only slightly adapted to the horse's back to being completely tailored. In order to create extreme examples with clearer implications, let us assume the latter case. At this highest degree of customization, almost no parts of the manufacturing processes, except materials, skills and journeys to the customers' stable, can be shared between different saddles since every product is unique. Customers proactively influence the marketing mix with regard to product policies because they require the saddler to measure horses and riders, and they define their needs regarding material, design, size and weight. Even the price, to a certain extent, may be negotiable and can be influenced by customers. The same applies to services after the final delivery of the saddle: it is the role of the customer to address the saddler if further adjustments are needed (e.g. changes in cushion or saddle tree). The saddler gathers personal information on individual preferences but customers actively disclose them. Saddlers react to the information they get and try to fulfill the customer's needs based on concrete orders.

Production and purchase prices are high compared to machine-made saddles from European brands. Extensive requirements for the saddler's knowledge, long production times because of hand-crafting and rare, hard-to-access materials are reasons for expensive production. This case is therefore substantially distinguished from mass customization where products are individualized in large quantities and cost levels are close to mass production (Piller, 1998). Practical examples where mass customization is profitably implemented are computers and printers from the firm "Hewlett Packard" (Feitzinger & Lee, 1996) and women's blue jeans by "Levi Strauss & Co." (Pitta, 1998).

Even though we are dealing with an extreme case of customization, this does not mean that we cannot find elements of personalization. Once a saddler has been informed about the customer's preferences and personal data, he will adjust his offer accordingly. Here, we find asymmetric information in so far as the saddler has an overview of different leathers and other materials and their prices. Further, he knows how long production will take. If, for example, a wealthy customer is faced with options with regard to material (especially additional product features) then a different price may be given than to less sound, more price-sensitive customers.[10] A saddler could also feel less pressure to deliver quickly if he assumes the customer is patient. In summary, saddlers can be assumed to discriminate among different offers based on the information particular customers have revealed. Price and product dimensions may be configured more or less opportunistically, depending on how the saddler evaluates his own bargaining power in each buyer-supplier relationship.[11] In the case of the above-described French saddle maker, supplier bargaining power is given in terms of market structure. The firm operates in an oligopolistic environment, which means that customer choices are strongly limited.

A saddler will furthermore use information on his customers to offer particular sale-services. Places of living, professions and preferences influence the types and locations of customer events, and the management of provided services. In particular, the time waiting until a saddler visits a customer's stable in order to take measurements or to execute after-sale adjustments follows the customer's switching costs. Particularly loyal customers may wait longer than critical ones. At this point, customers cannot influence processes; they are addressed by marketing policies according to the saddler's decision, whereas the latter uses information that customers disclosed consciously ex ante.

Thus, we find elements of personalization applied in our case. With reference to the definition by Arora et al (2008), one of the characteristics of personalization is that the supplier uses information about his customers' preferences to apply the classical marketing mix individually. This is done mainly with regard to product, price and communication. What remains different from the original definition is that the information base is not revealed by the initiative of the firm. Instead, customization can here be considered as an upstream process where the particular buyer reveals preferences and personal information in order to specify his demanded product.

In summary, the marketing for handcrafted, tailor-made saddles corresponds to customization in its strictest sense. We can reveal substantial differences between this and mass customization, which is often referred to in literature and describes the attempt to customize with mass-production efficiency. Instead, customized saddles are produced in small quantities with turn-around times lasting several months while production cost is high and the level of process standardization is low. Personalization strategies can also be identified. The saddler has an incentive to adjust his marketing policies regarding materials, prices and communication to-

---

[10] For asymmetric information in bilateral bargaining, see Samuelson (1984).

[11] For price discrimination, see Varian (1989).

ward information he has about his customers' preferences and socio-demographic data. Table 1 gives an overview over key results from this paragraph.

Table 1. Overview of key results

Dimensions of one-to-one strategies	Marketing for customized riding saddles	Mass customization	Personalization
Proactive part in data collection	Customer (saddler collects data reactively)	Customer	Firm (customers disclose information unconsciously)
Design of marketing policies	Both	Customer	Firm
Processes shared among products	Every saddle is individually handcrafted, materials and journeys to visit customers at their stable can be shared	Processes are shared as far as possible in order to keep unit cost low. Materials and standardized product bases are examples of common features.	Information technologies are shared in order to manage customer data.
Cost	High (high requirements on human assets, rare materials and long turnaround times)	The goal is to produce with mass-production efficiency (large quantities and low unit costs).	Cost depends on the efficiency of applied information technologies.

Theory development: Conceptualizing one-to-one luxury marketing

Hypothesis 1

So far, we have revealed customization and personalization strategies in our case. Furthermore, tailored high-quality riding saddles are found to possess typical characteristics of luxury goods. We have also identified major overlaps between the target groups of both branches in China, luxury good consumers and potential buyers of customized saddles. Not only socio-demographic data is comparable but riders also directly signaled preferences for quality, luxury and individualization. Thus, we interpret this as an example for the application of one-to-one-marketing for luxury goods in China. In order to generalize our findings for Chinese luxury industries, we have to address the question if one can extrapolate from equestrians to Chinese luxury consumers in general. To answer this, we additionally need to take the cultural context into consideration.

One-to-one-marketing involves strong relationships between manufacturers and particular customers (Peppers & Rogers, 1993). The marketing strategies are tailored to individual customers, since the manufacturer is focusing its activities to one single entity in the market. Once a product is customized it cannot be marketed to remaining customers any longer, or at least only with a substantial price reduction (customized riding saddles again represent an example, where a product is hardly transferable to other customers once it is fully adjusted). Thus, suppliers appear to be vulnerable if their opportunity to diversify risks is limited. While mass products without any adjustment to individual preferences can be sold to a large customer base, customized products are dependently manufactured for their principals. A supplier will thus reduce his risks by means of a contract.[12] The customer has to ensure *ex ante* that he will buy the product as ordered. Terms and conditions are fixed and both supplier and buyer are bound to the order they agreed on, making them interdependent. Customers are additionally vulnerable because they have to disclose private information to the supplier in order to have the product tailored to their preferences.

In summary, we find a personalized relationship that lasts at least for several months and is thus distinct from anonymous purchases in mass markets. Based on Confucian tradition, Chinese society is characterized by being built on strong interpersonal relationships. (Lu, 2008). The cultural framework of Hofstede (1993) identifies long-term orientation (versus short-term orientation) as one of six dimensions with which a culture can be defined. Relationships, in

---

[12] See for contracting in principal-agent relationships Holmstrom and Milgrom (1991).

private and business contexts, are meant to be stable and liable (Lee & Dawes, 2005). Harmony, attachment and predominant ties are characteristically for the Chinese phenomenon of *guanxi*. The latter describes interpersonal relations within social circles and their significant influence on every individual's rights, social behavior and business practices. Lee and Dawes (2005) state "Guanxi lies at the heart of China's social order, its economic structure, and its changing institutional landscape. It is considered important in almost every realm of life, from politics to business and from officialdom to street life." (p. 29). In accordance with *guanxi*, loyalty and serving the community's needs are placed above personal concerns (Wong & Ahuvia, 1998).

Thus, China represents an environment where customers favor personalized purchase processes and trust-based, long-term orientated (formal or informal) relationships with suppliers. Luxury (customized) products in particular may be preferably bought in personal, trustworthy settings as their perceived value is strongly connected with purchase experience and well-being (Addis & Holbrook, 2001; Yeoman & Mc Mahon-Beattie, 2006).

Apart from interpersonal relationships, we refer to privacy protection as a second crucial concern in the debate of personalization strategies (Wind & Rangaswamy, 2001). In China, rules for privacy protection are less comprehensive, while the private sphere is generally less valued than in Western cultures. This can again be explained by Chinese Confucian values, whereby the needs of the community (like family) dominate individual rights (Hofstede, 1993). Nevertheless, researchers are observing a trend toward strengthening privacy and data protection, and the individual is increasingly being seen as independent from the larger society and the state. This enforcement of individual rights has been enhanced since China has opened up to Western values and traditions (Ess, 2006; Ess, 2005; Yao-Huai, 2005).

Despite recent developments, it can still be assumed that Confucian values are advantageous in terms of information seeking. Chinese people are likely to reveal their preferences more comprehensively because they may be less concerned with privacy protection than customers in individualistic cultures. Our findings lead us to derive the following first hypothesis:

H1: One-to-one marketing is effectively applicable to Chinese luxury consumers.

Hypothesis 2

Where high-quality, handcrafted leather saddles are customized, every single rider and his or her horse is the focus of a manufacturer's communication. Information about body size and the shape of the horse's back, about riding styles and associated individual demands, and, in extreme cases, about the riders' physical problems are used to correctly fit the saddle. Sale services and other specific marketing actions are developed to encourage the customer to pay a price that is far above the average price for mass-produced riding saddles. In conclusion, we have a case that closely links one-to-one marketing to luxury.

Other examples where luxury is marketed using one-to-one marketing strategies are luxury tourism (Murphy, Schegg, & Olaru, 2007), jewelry (Ahde, 2007), cars (Pitta, 1998) and private jets (Osterwalder & Pigneur, 2003). Pine II (1993) as well as Addis and Holbrook (2001) looked into marketing luxury with the help of customization strategies from the theoretical point of view over a decade ago. They state that luxury products are predestinated to become customized because they are unique, distinctive, higher priced and connected with hedonic consumption experience. Broekhuizen and Alsem (2002) propose a conceptual model where success factors of the external and internal organizational environment are considered in order to evaluate if mass customization is advantageous for a particular firm. They refer to the luxury value as one of four different crucial product factors and hypothesize that the luxury level positively influences the probability that customization turns into success.

If we compare *one-to-one luxury marketing* (OLM), that should here define the combination of one-to-one marketing and luxury marketing, with pure forms of classical marketing strategies we can find distinct characteristics concerning price level and degree of individualization. While mass marketing is associated with low prices and the highest degree of standardization, OLM presents quite the opposite. Prices and possibilities of customization are among their highest in the saddle market. Classical luxury marketing as used by prestigious fashion, jewelry, housing or entertainment brands faces high prices but has more or less standardized products. One-to-one marketing can instead be categorized as a high level of individualization but is per definitionem not necessarily associated with high prices because it refers to mass customization according to existing literature (Arora et al. 2008, Peppers, Rogers, & Dorf, 1999). Figure 9 displays what we found.

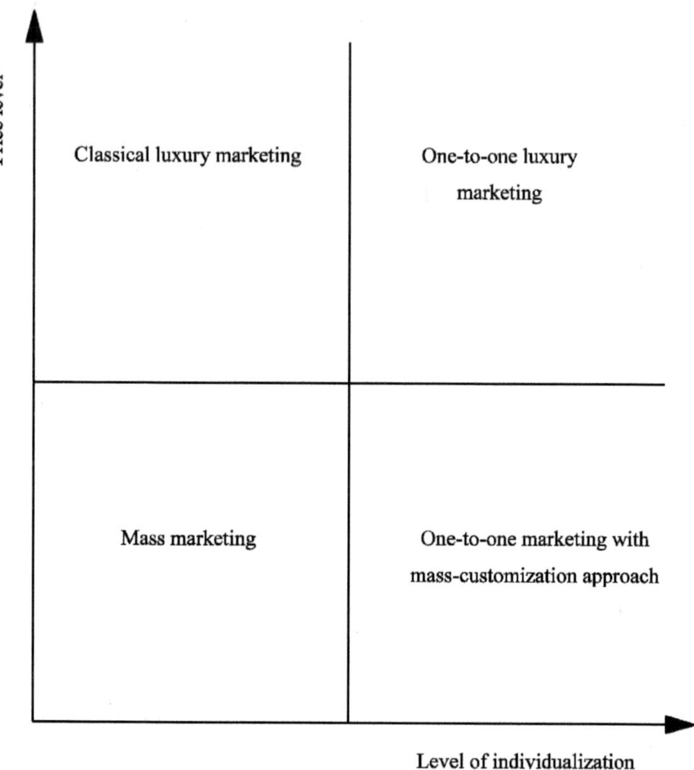

Figure 9. Classification of one-to-one luxury marketing

OLM appears to fill the conceptual gap between luxury marketing and one-to-one-marketing. It refers to fully individualized products on higher price levels, products where customers are willing to pay mark-ups in order to have their product fully tailored to their preferences. Market observations show that luxury consumers demand customization. Uniqueness and "striving for the special something" are generally crucial consumption motives for luxury goods (Kisabaka, 2001; Lu, 2008; Vigneron & Johnson, 2004; Wiedmann, Hennigs, & Siebels, 2009). We find customers' preferences to be compatible with product individualization. Thus, creating a framework that is built upon one-to-one relationships between firms and consumers may provide major contributions to luxury marketing. From our results, we deduce our second hypothesis:

86

H2: One-to-one luxury marketing combines the highest product price levels with the highest levels of product individualization.

Hypothesis 3

There are common motives for luxury consumption found in different cultural settings, and the fact that Chinese riders not only demand customized saddles but in particular by European riders allows us to generalize at this point. Explanations can be found in recent trends in consumption patterns within both cultures.

First, as China opens up politically, its society, especially young and middle-aged people, are shifting toward Western consumption patterns. Chevalier and Lu (2010) find that in recent decades, as wealth, modernization and openness to foreign markets has increased, Chinese *luxury lovers*, which are mainly young, well-educated people, have adopted Western motives for luxury consumption. Ess (2005), Lu (2008) and Yao-Huai (2005) also state that Western-style values influence behavioral patterns in China and are partially replacing traditional *Confucian* values. The same applies to individualism versus collectivism. With Western orientation, especially young and middle-aged people are becoming more independent and individualistic. Consistent with this, our face-to-face interviews revealed that Chinese riders often chose American or European universities for their studies, or for the studies of their children. Getting an international education is highly valued in this social environment, and Western countries were assumed better at providing this. The same applies to their training in equestrian sports. Parents are willing to spend more for their children to be trained in European equestrian nations, like Germany, in order for them to become more skilled in riding than would be possible in their home country.

Thus, we find that our target group is more driven by modern consumption motives like hedonism and individualization, and less influenced by traditional Chinese values like social independency, hierarchy and patriotism.

Second, in Western countries, consumers have recently started to value consumption experience and subjectivity more. The hedonic component of luxury goods has become a new focus and has pushed old motives like prestige-seeking and conspicuousness into the background. Especially in case of luxury consumption, self-concepts should be realized and the individualism of a person's lifestyle should be experienced (Addis & Holbrook, 2001; Ascheberg, Meurer, & Oesterling, 2012; Atwal & Williams, 2009; Hagtvedt & Patrick, 2009; Yeoman,

2011). We transfer this development to our case and assume that product individualization marketing strategies satisfy not only East Asian but also Western luxury consumers. Accordingly, we state our third hypothesis as follows:

H3: The findings from the Chinese horse sports case can be transferred to luxury marketing in other cultural contexts. One-to-one luxury marketing is effectively applicable to transcultural luxury consumers.

Summary and remarks on future research

From our case study, we derive three hypotheses dealing with the integration of two existing marketing strategies. We refer to this integration as *one-to-one luxury marketing* (OLM), which is defined as a mixture of one-to-one marketing and classical luxury marketing. It can be visualized by using a chart diagram with two axes—price level and level of individualization—where OLM is categorized in the upper right hand corner. It refers to fully individualized marketing strategies for luxury products in the highest quality categories. In our case, we focused on customized handcrafted leather riding saddles sold in China, which are valued at the highest level in the respective market.

Results from our theoretical and empirical considerations according to target group characteristics and environmental factors have shown that the case lets us conclude for luxury customers in general. Accordingly, OLM is hypothesized to satisfy the luxury customer's demand for customization, personalized sales services and individual marketing. It may effectively be applied as an instrument to serve the market for luxury at high price levels and low quantities. Even if our study is limited to the Chinese setting, whose culture is influenced by Confucian values and norms, reasons can be found for applying our findings to Western cultural environments.

This hypothesis should be tested through further data. First, additional Chinese manufacturers for luxury goods other than equestrian equipment should be investigated. Single or multiple case studies and empirical considerations of market performance and financial data would provide information on our first hypothesis. In terms of our second hypothesis, research should evaluate if OLM can be legitimized as a new standard for luxury marketing. Multiple-case studies would be helpful to address the demand for theoretical considerations to the connection of one-to-one concepts and luxury. Finally, cross-border market performance and financial data of several luxury manufacturers that apply OLM would serve as a base for testing the third hypothesis.

If further research could validate our hypotheses or, at least the first and the second, the findings from this case may provide a significant contribution to luxury marketing. Despite the fact that there has been a growing global trend for enhanced luxury goods sales recently, research in the field of high-end markets is limited. At the same time, we observed major interest from companies, at least in the branch of luxury riding equipment, to conceptualize of adequate strategies to serve the growing market.

The explanatory power of our case study is constrained by the small sample size in the questionnaire of 67 participants, assumable biases based on social desirability associated with evaluating consumer attitude towards luxury and the non-integration of market performance and financial data from customized saddle sales in China into our analysis. Since the firm we investigated just started to serve the Chinese market, we were not able to generate information on profitability.

References

Addis, M., & Holbrook, M. (2001). On the conceptual link between mass customisation and experiential consumption: An explosion of subjectivity. *Journal of Consumer Behavior, 1*(1), 50-66.

Ahde, P. (2007, August): Appropriation by adornments: personalization makes the everyday life more pleasant, in: Proceedings of the 2007 conference on Designing pleasurable products and interfaces, pp. 148-157. ACM.

Albrecht, C. M., Backhaus, C., Gurzki, H., & Woisetschläger, D. M. (2013). Drivers of brand extension success: What really matters for luxury brands. *Psychology & Marketing, 30*(8), 647-659.

Arora, N., Dreze, X., Ghose, A., Hess, J. D., Iyengar, R., Jing, B., ..., & Zhang, Z. J. (2008). Putting one-to-one marketing to work: Personalization, customization, and choice. *Marketing Letters, 19*(3-4), 305-321.

Ascheberg, C., Meurer, J., & Oesterling, A. (2012). The Luxury Universe–Angebots-und Kundensegmentierung globaler Luxusmärkte als Basis für erfolgreiche Positionierungsstrategien. In C. Burmann, V. König, & J. Meurer (Hrsg.), *Identitätsbasierte Luxusmarkenführung* (S. 85-101). Springer Fachmedien Wiesbaden.

Atsmon, Y., Ducarme, D., Magni, M., & Wu, C. (2012). The McKinsey Chinese Luxury Consumer Survey *Luxury Without Borders: China's New Class of Shoppers Take on the World*. McKinsey & Company (Ed.).

Atsmon, Y., Magni, M., Li, L., & Liao, W. (2012). *Meet the 2020 Chinese Consumer. McKinsey Consumer & Shopper Insights*. McKinsey & Company.

Atwal, G., & Williams, A. (2009). Luxury brand marketing–The experience is everything!. *Journal of Brand Management, 16*(5), 338-346.

Berger, C., & Piller, F. (2003). Customers as co-designers. *Manufacturing Engineer-London, 82*(4), 42-45.

Broekhuizen, T. L. J., & Alsem, K. J. (2002). Success Factors for Mass Customization: A Conceptual Model. *Journal of Market-Focused Management, 5*(4), 309-330.

Chevalier, M. & Lu, P. (2010). Luxury China: Market Opportunities and Potential, John Wiley&Sons (Asia) Limited, Singapore.

CLSA Asia Pacific Markets (Ed.) (2011). *Dipped in Gold: Luxury lifestyles in China/HK*. Special consumer report.

Cocq, P. D., Weeren, P. V., & Back, W. (2004). Effects of girth, saddle and weight on movements of the horse. *Equine veterinary journal, 36*(8), 758-763.

D. R. Vereinigung, D. R. (FN) (2001). Faszination Zukunft-Neue Perspektiven im Pferdesport. *Marktanalyse Pferdesportler in Deutschland.*

D´arpizio, C. (2014). *Luxury Goods Worldwide Market Study Spring 2014.* Bain and Company (Ed.), Boston.

Danziger, P. (2005). *Let them eat cake: Marketing luxury to the masses-as well as the classes.* Kaplan Publishing.

Danziger, P. N. (2005). Let Them Eat Cake: Marketing Luxury to the Masses – As Well as the Classes. Dearborn Trading, New York.

Darke, P., Shanks, G., & Broadbent, M. (1998). Successfully completing case study research: combining rigour, relevance and pragmatism. *Information systems journal, 8*(4), 273-289.

Degen, R. J. (2009). Opportunity for luxury brands in China. *Journal of Brand Management, 6*(3/4), 75-85.

Dubois, B., & Laurent, G. (1994). Attitudes toward the concept of luxury: An exploratory analysis. *Asia-Pacific Advances in Consumer Research, 1*(2), 273-278.

Dubois, B., & Laurent, G. (1995). Luxury possessions and practices: an empirical scale. *European Advances in Consumer Research, 2*, 69-77.

Dubois, B., & Paternault, C. (1995). Observations: Understanding the world of international luxury brands: The" dream formula.". *Journal of Advertising Research.*

Dyson, S. J. (2000). Lameness and poor performance in the sports horse: dressage, show jumping and horse trials (eventing). *Proc. Am. Vet. Med. Assoc., 46,* 308-315.

Eckjans, S., & Hartmann, L. (2013). Neue Märkte im Pferdesektor–China. *Goettinger Pferdetage '13: Zucht, Haltung und Ernährung von Sportpferden. Tagungsband Goettinger Pferdetage 2013.* 33-35.

Eisenhardt, K. M. (1989). Building theories from case study research. *Academy of management review,* 14(4), 532-550.

Ess, C. (2005). "Lost in Translation"?: Intercultural Dialogues on Privacy and Information Ethics (Introduction to Special Issue on Privacy and Data Privacy Protection in Asia). *Ethics and Information Technology,* 7(1), 1-6.

Ess, C. (2006). Ethical pluralism and global information ethics. *Ethics and Information Technology,* 8(4), 215-226.

Fassnacht, M., Kluge, P. N., & Mohr, H. (2013): Pricing Luxury Brands: Specificities, Conceptualization and Performance Impact. *Marketing ZFP – Journal of Research and Management,* 35(2), 104-117.

FEI Database (2013). Competitors. Last accessed: 22 October 2013.

Feitzinger, E., & Lee, H. L. (1997). Mass customization at Hewlett-Packard: the power of postponement. *Harvard Business Review, 75,* 116-123.

Flyvbjerg, B. (2006). Five misunderstandings about case-study research. *Qualitative inquiry,* 12(2), 219-245.

Focus (2009). *Der Markt für Luxusgüter.* Projekt Zukunft. URL: www.medialine.de/.../marktanalysen/.../foc_ma_luxusgueter_200908.pdf. Last accessed: 06 February 2015.

Gabel, M. (2013). *Schön, edel, prunkvoll: Die französische Erfolgsgeschichte der Luxusindustrie.* Dossier. Bundeszentrale für politische Bildung (Hrsg.). URL: http://www.bpb.de/internationales/europa/frankreich/152660/luxusindustrie. Last accessed: 09 February 2015.

Garner, J. (2005). *The Rise of the Chinese Consumer – Theory and Evidence.* Credit Suisse First Boston (Europe) Limited, John Wiley&Sons Limited, West Sussex.

Gille, C., Hoischen-Taubner, S., & Spiller, A. (2011). Neue Reitsportmotive jenseits des klassischen Turniersports. *Sportwissenschaft, 41*(1), 34-43.

Hagtvedt, H., & Patrick, V. (2009). The broad embrace of luxury: Hedonic potential as a driver of brand extendibility. *Journal of Consumer Psychology, 19*, 608-618.

Hellhammer, S. (2009). *Marketingforschung im Luxussegment: Mechanismen, Märkte und Methoden* (Doctoral dissertation, Otto-von-Guericke-Universität Magdeburg, Magdeburg).

Hofstede, G. (1993). Cultural constraints in management theories. *The Academy of Management Executive, 7*(1), 81-94.

Holmstrom, B., & Milgrom, P. (1991). Multitask principal-agent analyses: Incentive contracts, asset ownership, and job design. *Journal of Law, Economics, & Organization, 7*, 24-52.

Imoersteg, M. (1967). *Die Entwicklung des Konsums mit zunehmendem Wohlstand.* Volkswirtschaftlich-wirtschaftsgeographische Reihe – Band 16, Polygraphischer Verlag AG Zürich und St. Gallen.

Kapferer, J. N., & Bastien, V. (2009). The specificity of luxury management: Turning marketing upside down. *Journal of Brand Management, 16*(5), 311-322.

Kisabaka, L. (2001). *Marketing für Luxusprodukte* (Vol. 32). epubli.

KPMG (2008) (Ed.). China's Luxury Consumers: Moving up the curve. Consumer markets. URL: *http://www.kpmg.com/cn/en/issuesandinsights/articlespublications/pages/luxury-consumers-200804.aspx.* Last accessed: 06 February 2015.

Lee, D. Y., & Dawes, P. L. (2005). Guanxi, trust, and long-term orientation in Chinese business markets. *Journal of international marketing, 13*(2), 28-56.

Liang, T. P., Chen, H. Y., & Turban, E. (2009, August). Effect of personalization on the perceived usefulness of online customer services: A dual-core theory. In *Proceedings of the 11th International Conference on Electronic Commerce* (pp. 279-288). ACM.

Lu, P. X. & Pras, B. (2011). Profiling Mass Affluent Luxury Goods Consumers in China: A Psychographic Approach. *Thunderbird International Business Review, 53*(4), 435-455.

Lu, P. X. (2008). *Elite China: Luxury Consumer Behavior in China.* John Wiley&Sons (Asia) Limited, Singapore.

Mason, R. (1993). Cross-cultural influences on the demand for status goods. *European Advances in Consumer Research, 1*, 46-51.

Murphy, J., Schegg, R., & Olaru, D. (2007). Quality clusters: dimensions of email responses by luxury hotels. *International Journal of Hospitality Management, 26*(3), 743-747.

Osterwalder, A. & Pigneur, Y. (2003, September). Modeling value propositions in e-Business, in: Proceedings of the 5th international conference on Electronic commerce, pp. 429-436. ACM.

Peppers, D., & Rogers, M. (1997). *The one to one future*. New York: Doubleday.

Peppers, D., Rogers, M. & Dorf, B. (1999). Is your company ready for One-to-one marketing. *Harvard Business Review*, 77(1), 151-160.

Peppers, D., Rogers, M. (1993). *The One to One Future: Building Relationships One Customer at a Time*. New York: Currency/Doubleday.

Pettigrew, A. (1988). Longitudinal field research on change: Theory and practice. *Organization science, 1*(3), 267-292.

Pflanz, C. (2003). Faszination Luxus. In C. Pflanz (Hrsg.), *Faszination* (S. 90-97). Gabler Verlag.

Piller, F. T. (1998). *Kundenindividuelle Massenproduktion: Die Wettbewerbsstrategie der Zukunft*. Hanser.

Pine II, B. J. (1993). Making mass customization happen: strategies for the new competitive realities. *Strategy & Leadership, 21*(5), 23-24.

Pitta, D. A. (1998). Marketing one-to-one and its dependence on knowledge discovery in databases. *Journal of Consumer Marketing, 15*(5), 468-480.

Roland Berger Strategy Consultants (Ed.) (2012). *A brand awareness upgrade: Welcoming a new era in the Chinese luxury market*. think:act STUDY.

Samuelson, W. (1984). Bargaining under asymmetric information. *Econometrica: Journal of the Econometric Society, 52*(4), 995-1005.

Scheer, C., & Loos, P. (2002, September). Customer-oriented products and services–Classification, discussion of traditional concepts and suggestion of an internet-based business model. In *Proceedings of the 3rd International ICSC Symposium on Engineering of Intelligent Systems (EIS), Workshop on Information Systems for Mass Customization (ISMC)*.

Smith, W. R. (1956). Product differentiation and market segmentation as alternative marketing strategies. *The Journal of Marketing, 21*(1), 3-8.

Tsai, S. P. (2005). Impact of personal orientation on luxury-brand purchase value. *International Journal of Market Research, 47*(4), 429-454.

Varian, H. R. (1989). Price discrimination. *Handbook of industrial organization, 1*, 597-654.

Veblen, T. (1899). *The theory of the leisure class*. Oxford University Press.

Vigneron, F., & Johnson, L. W. (1999). A review and a conceptual framework of prestige-seeking consumer behavior. *Academy of Marketing Science Review, 1*(1), 1-15.

Vigneron, F., & Johnson, L. W. (2004). Measuring perceptions of brand luxury. *The Journal of Brand Management, 11*(6), 484-506.

Wiedmann, K. P., Hennigs, N., & Siebels, A. (2009). Value-based segmentation of luxury consumption behavior. *Psychology & Marketing, 26*(7), 625-651.

Wind, J., & Rangaswamy, A. (2001). Customerization: the next revolution in mass customization. *Journal of interactive marketing, 15*(1), 13-32.

Wong, N. Y., & Ahuvia, A. C. (1998). Personal taste and family face: Luxury consumption in Confucian and Western societies. *Psychology and Marketing, 15*(5), 423-441.

Yao-Huai, L. (2005). Privacy and data privacy issues in contemporary China. *Ethics and Information Technology, 7*(1), 7-15.

Yeoman, I. & McMahon-Beattie, U. (2010). The changing meaning of luxury. In: I. Yeoman and U. McMahon-Beattie (eds.) *Revenue Management: A Practical Pricing Perspective*, Chapter 6, Basingstoke, UK: Palgrave MacMillan, pp. 62–85.

Yeoman, I. (2011). The changing behaviours of luxury consumption. *Journal of Revenue & Pricing Management, 10*(1), 47-50.

Yeoman, I., & McMahon-Beattie, U. (2006). Luxury markets and premium pricing. *Journal of Revenue and Pricing Management, 4*(4), 319-328.

Zhou, L., & Hui, M. K. (2003). Symbolic value of foreign products in the People's Republic of China. *Journal of International Marketing, 11*(2), 36-58.

Chapter II: Luxury Food: Definition and Consumption Motives

II.1   **Luxusmarketing bei Lebensmitteln: Eine empirische Studie zu Dimensionen des Luxuskonsums in Deutschland**

Autoren: **Thea Schneider, Laura Hartmann, Achim Spiller**

Georg-August-Universität Göttingen

Dieser Beitrag wurde in ähnlicher Form als Diskussionspapier der Georg-August-Universität Göttingen veröffentlicht (Diskussionspapier 1502, *Diskussionspapiere der Georg-August-Universität Göttingen, Department für Agrarökonomie und Rurale Entwicklung*, ISSN 1865-2697).

Zusammenfassung

Heutzutage zeigt sich das Ernährungsverhalten der deutschen Konsumenten immer differenzierter und ist mit einem wachsenden Involvement beim Kauf von Lebensmitteln verbunden. In diesem Zusammenhang steigt die Nachfrage nach Qualitätslebensmitteln und Premium-Marken. Für das Marketing im Bereich hochpreisiger Lebensmittel ergibt sich die Frage nach den kaufmotivierenden Produkteigenschaften und der Abgrenzung zum Konsumentenverhalten auf dem allgemeinen Lebensmittelmarkt. Die Konzeptionierung zielgruppenspezifischer Marketingmaßnahmen erfordert die Kenntnis der Einstellungen und Einflussfaktoren von Konsumenten der sogenannten Luxus-Lebensmittel. Bislang sind unseres Wissens hierzu jedoch kaum empirische Studien vorhanden. Basierend auf einer explorativen Faktorenanalyse untersucht die vorliegende Studie die Erwartungen von Konsumenten an hochpreisige Lebensmittel und dem gegenübergestellt, die kaufbeeinflussenden Faktoren auf dem allgemeinen Lebensmittelmarkt. Die Ergebnisse zeigen, dass Konsumenten sich vom Kauf hochpreisiger Lebensmittel im Wesentlichen einen Zugewinn an Nachhaltigkeit, Genuss und Geschmack versprechen. Beim Kauf von normalpreisigen Lebensmitteln lassen sich Konsumenten hingegen eher von den Faktoren Marketing und Preis-Leistungsverhältnis leiten.

Schlüsselwörter

Hochpreisige Lebensmittel – Konsumenten – Erwartungen – Luxusmarketing

Abstract

Nowadays, the nutritional behavior of German consumers appears to become more differentiated and is moreover related to a high involvement during the purchase of food. As a result, the demand for high quality food products has increased. The question arises which consumption motives are significant in the market for high-price food products and how these differ in comparison to the conventional food market. For the conception of target-group specific marketing strategies, new empirical consumer studies on the market for this so-called luxury food are needed. The developments on the German food market have a high topicality, but there is a lack of those investigations. Based on an explorative factor analysis, this study investigates the expectations of German consumers toward purchasing upscale food products. Moreover, it compares this with factors that influence purchase decisions in the general food market. The results of the analysis show that consumers of upscale food products especially expect gains in sustainability, indulgence and taste, whereas in the market for general food products, consumers are more influenced by marketing and price-performance ratio.

Keywords

Upscale Foods, Consumers, Expectations, Luxury Marketing

Einleitung

Der Luxusmarkt ist ein Markt, der sich im Wachstum befindet. In den vergangenen 20 Jahren hat sich die Zahl der Konsumenten von Luxusartikeln weltweit mehr als verdreifacht. In Zahlen gesprochen, heißt dies, dass die Anzahl der Luxuskäufer von rund 90 Millionen im Jahr 1995 auf 330 Millionen Ende 2013 gestiegen ist. Des Weiteren treten jedes Jahr zehn Millionen neue Kunden auf den Markt für Luxuswaren (D´arpizio & Levato, 2014). Laut Yeoman und McMahon-Beattie (2006) liegt der Grund hierfür in den steigenden Einkommen der Verbraucher, die zu einer Erhöhung ihrer Ansprüche führen.

Luxus wächst nicht nur global, sondern hat ebenfalls Konjunktur in der Bundesrepublik Deutschland (BRD), denn „die deutschen Konsumenten werden luxusaffiner und legen ihre Luxusaversionen ab" (Meurer, 2012). Des Weiteren thematisiert Meurer (2012) in diesem Kontext einen selbstverstärkenden Prozess, da sich das Luxusimage durch eine zunehmende Prägung von Nachhaltigkeit und Verantwortung in der Gesellschaft verbessert.

Der Begriff des Luxus erstreckt sich laut Felder (1997) über eine ökonomische, soziale, politische, anthropologische, theologische und ethische Dimension. Grugel-Pannier (1996) definiert den Begriff des Luxus als „die durchschnittliche Lebenshaltung weit überschreitenden Aufwand" oder als „die ausschweifende Lebensweise von Menschen, deren Sinnengenuss den ‚normalen' oder ‚richtigen' Standard übersteigt".

Auf dem deutschen Markt für Lebensmittel ist ein Wandel zu beobachten, durch den sich der Stellenwert von Produktqualität, Nachhaltigkeit, Authentizität sowie von hedonistischen Werten wie Genuss aus Konsumentensicht erhöht (Nestlé Deutschland AG, 2012). Das Kriterium eines niedrigen Preises hingegen verliert zunehmend an Bedeutung für die Kaufentscheidung (Nestlé Deutschland AG, 2012; SGS, 2014). Damit korrespondierend beschreiben Kirig und Rützler (2007) einen Lebensmitteltrend in Deutschland, der die herkömmliche Geiz-Mentalität in ein erhöhtes Gesundheits- und Qualitätsbewusstsein, in die Sehnsucht nach hedonistischem Erleben und Luxus wandelt. Diese Entwicklungen motivieren die empirische Forschung im Bereich so genannter Luxus-Lebensmittel.

Van der Veen (2003) bezeichnet diese als Lebensmittel, die einer Verfeinerung der Beschaffenheit, des Geschmacks, des Fettanteils und anderer Qualitätsparameter dienen. Sie können sich entweder durch Qualität oder Quantität von anderen Lebensmitteln abgrenzen. Jedoch ist Bezug nehmend auf den Luxusmarkt für Lebensmittel feststellbar, dass hier keine allgemein gültige Begriffsdefinition existiert. Aus diesem Grund ist es wichtig, sich dem Konstrukt auf andere Weise zu nähern, um die wesentlichen Eigenschaften von Luxuslebensmitteln aus

Verbrauchersicht benennen zu können und um Unterschiede zu „normalen" Lebensmitteln zu ermitteln.

Im vorliegenden Beitrag werden Luxus-Lebensmittel als hochpreisige Lebensmittel oder auch Feinkost/Delikatessen von der Angebotsseite her anhand des Preises definiert und beziehen sich folglich auf ein betriebswirtschaftliches Luxusverständnis. Dubois, Laurent, und Czellar (2001) bezeichnen den Preis als eines von sechs maßgeblichen Charakteristika für Luxusgüter. Außerdem kann dieser im Konsumentenverhalten als Moderatorvariable beschrieben werden, die dem Verbraucher vermittelt, dass eine bessere Produktqualität vorliegt (Wiedmann, Hennigs, & Siebels, 2007).

Die vorliegende Studie hat zum Ziel, die Erwartungen von Verbrauchern an hochpreisige Lebensmittel zu ermitteln. Des Weiteren sollen Zusammenhänge mit dem allgemeinen Lebensmittelmarkt beleuchtet werden. Die Ergebnisse der Analysen dienen dazu, zukünftige Marketingvorhaben in dem Sektor für hochpreisige Lebensmittel zu stützen. Für die Forschung im Bereich des Luxusgütermarketings leisten die Ergebnisse einen Beitrag zur Analyse von Konsummotiven auf einem sich neu etablierenden Luxusmarkt. Das Phänomen des sich segmentweise zunehmend als Luxusmarkt definierenden Lebensmittelmarktes zeigt D´arpizio (2014).

Dimensionen des Luxuskonsums

Der Begriff des Luxus ist laut Van der Veen (2003) selten voll umfassend definiert, und seine Definitionen variieren in der Schwerpunktsetzung auf ökonomischen und sozialen Aspekten. Laut Lasslop (2005) wird Luxus in Bezug auf die ökonomische Perspektive auf der einen Seite objektorientiert zur Abgrenzung von Gütern und Marken diskutiert und auf der anderen Seite verhaltensorientiert hinsichtlich deren Konsum. Heutzutage gilt es, die Rolle der Luxusgüter im Kontext der Umwelt des Konsumenten zu verstehen und die für die Kaufentscheidung relevanten Motive nachzuvollziehen (Hellhammer, 2009).

Dubois, Laurent und Czellar (2001) entwickelten einen neuen Ansatz, um das Luxuskonzept seitens der Konsumenteneinstellungen zu bewerten. In ihrer Forschungsarbeit identifizierten die Autoren insgesamt sechs Faktoren, die als maßgebliche Charakteristika von Luxusgütern gelten. Als Faktoren seien hier eine überdurchschnittlich hohe Produktqualität, ein sehr hoher Preis, Einzigartigkeit, Ästhetik, Historie und Nicht-Notwendigkeit genannt (Dubois, Laurent & Czellar, 2001). Laut Lasslop (2005) schaffen diese spezifischen Wahrnehmungsmuster einen herausragenden ideellen Nutzen für den Nachfrager. Folglich zeigen sich in der Konsumentenwahrnehmung von Luxus voneinander unabhängige Dimensionen. Dementsprechend stellen sich die Beziehungen der Konsumenten zu Luxus als multidimensional dar (Dubois, Laurent, und Czellar, 2001).

Wiedmann, Hennigs und Siebels (2009) veranschaulichen weiterführend diese Multidimensionalität in einem Modell, das relevante kognitive und emotionale Luxusdeterminanten zu vier Dimensionen – finanzieller, funktionaler, individueller und sozialer Wert – vereint. Werte seien an dieser Stelle definiert als stark verfestigte Einstellungen, die für das eigene persönliche Leben relevant sind, beziehungsweise als präskriptive Erwartungen, die an die Gesellschaft gestellt werden (Kröber-Riel & Gröppel-Klein, 2013). Wie in Abbildung 1 gezeigt, werden die vier Dimensionen durch insgesamt neun Faktoren präzisiert, um letztendlich die individuelle Wahrnehmung von Luxus ableiten zu können.

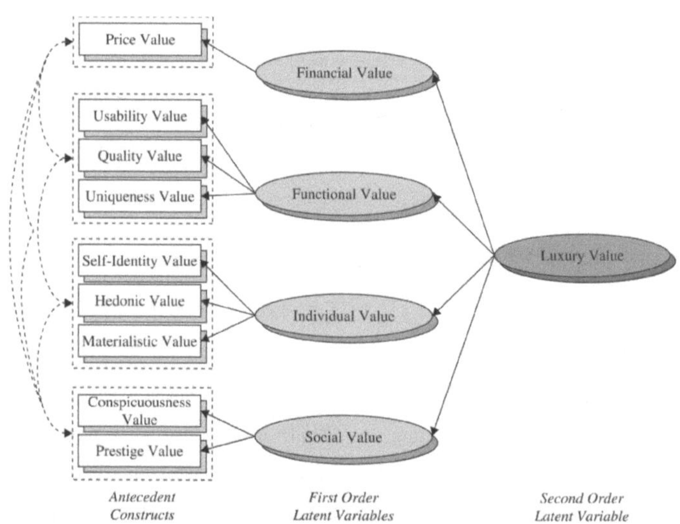

Abbildung 1. Dimensionen des Luxury Value[13]

Wiedmann, Hennigs und Siebels (2007, 2009) stellen sich die Frage nach den individuellen Einstellungen von Konsumenten in Bezug auf die charakteristischen Eigenschaften eines Luxusproduktes. Die Autoren aggregieren die subjektiven Variablen (antecedent constructs) in

---

[13] Quelle: Wiedmann, Hennigs, and Siebels (2007, 2009)

vier nutzenbasierte Dimensionen (latent variables). Bei der ersten Dimension handelt es sich um den finanziellen Wert (financial value), der die monetären Aspekte des Luxuskonsums, Bezug nehmend auf den aktuellen Preis dieser Produkte, beinhaltet (Hanzaee, Teimourpour, & Teimourpour, 2013). Hierbei ist zu erwähnen, dass die Variable Preis auch im Hinblick auf die zweite Dimension (functional value) und die vierte Dimension (social value) als Moderatorvariable für den Qualitäts- und Prestigewert fungiert und dies somit exemplarisch bestehende Schlüsselzusammenhänge zwischen den unterschiedlichen Variablen und Dimensionen veranschaulicht (Wiedmann, Hennigs & Siebels, 2007, 2009). Der funktionelle Wert (functional value) fokussiert insgesamt drei Variablen. Er beruht einmal auf der grundlegenden Zweckmäßigkeit und dem eigentlichen Nutzen sowie auf der wahrgenommenen Einzigartigkeit und der Qualität des Luxusproduktes (Hanzaee, Teimourpour & Teimourpour, 2013; Wiedmann, Hennigs & Siebels, 2007, 2009).

In der dritten Dimension wird der individuelle Wert (individual value) ebenfalls anhand dreier Variablen ermittelt, welche sich auf die persönliche Einstellung des Konsumenten gegenüber Luxuskonsum beziehen. Ferner betrachtet die vierte Dimension den sozialen Wert des Luxuskonsums (social value). Dieser beschreibt den individuellen erkennbaren Nutzen, wenn das Produkt innerhalb einer sozialen Gruppe verwendet wird, und ist mit Geltungskonsum und Prestige verbunden (Wiedmann, Hennigs, & Siebels, 2007, 2009). Nach Wiedmann, Hennigs und Siebels (2007, 2009) lässt sich zusammenfassend sagen, dass alle vier nutzenbedingten Dimensionen des Luxuskonsums sich als voneinander unabhängig darstellen, jedoch stark miteinander interagieren und die individuelle Auffassung des Luxuswertes beeinflussen. Bezug nehmend auf den Lebensmittelmarkt stellen Wertehaltungen auch hier wichtige Determinanten dar, die den Lebensmittelkonsum beeinflussen und steuern (Nitzko & Spiller, 2014). Laut Prüne (2012) nehmen Nachhaltigkeitsbewusstsein und sozial-ökologisch korrektes Verhalten zunehmend einen zentralen Stellenwert im alltäglichen Leben der Gesellschaft ein. So zeigt sich, dass der Markt für Lebensmittel aus ökologischem Anbau weltweit wächst (Padilla Bravo et al., 2013). Ebenfalls kommt die Nationale Verzehrsstudie II (NVS II) zu dem Ergebnis, dass ökologische Erzeugung, Bioprodukte und Biosiegel für 38,8% der Befragten in der Bundesrepublik wichtige Punkte bei der Einkaufsentscheidung sind (Max Rubner-Institut, 2008). Dieser Wandel ist auch auf dem Markt für generelle Luxusgüter beobachtbar, denn parallel zum konventionellen Markt formiert sich auch in der Nachhaltigkeitsnische ein Luxussegment (Prüne, 2012). Nach Meurer (2012) liegt in der BRD eine Veränderung der Luxuswahrnehmung vor, denn das Verständnis von Luxus hat sich hin zu einem „Green Luxury" entwickelt.

Im Hinblick auf einen erkennbaren Wandel im Luxussegment belegen Kröber-Riel und Gröppel-Klein (2013), dass sich die Relevanz einzelner Werte im Zeitverlauf durchaus verschieben kann. Kisabaka (2001) thematisiert an dieser Stelle einen Wertewandel, der aus gegenwärtiger Sicht das Konsumentenverhalten vor allem durch drei Tendenzen beeinflusst: die Hedonisierung, die Sublimierung und die Individualisierung. Bei der hier zunehmenden Bedeutung von so genannten Selbstentfaltungswerten kann laut Kisabaka (2001) davon ausgegangen werden, dass der Konsum von Luxusprodukten mit hoher Wahrscheinlichkeit steigt. Dies kann für den Zeitraum bis heute bestätigt werden (D´arpizio & Levato, 2014).

Resümierend ist das Marketing auf Luxusgütermärkten aufgrund erweiterter Wertehaltung der Verbraucher neuen Dimensionen des Luxuskonsums und somit neuen Herausforderungen ausgesetzt. Die Aufgabe folgender Forschungsvorhaben sowie absatzpolitischer Maßnahmen auf dem Lebensmittelmarkt besteht darin, den neuen Konsummustern auf Luxusmärkten in Zukunft mehr Beachtung zu schenken. Die Dringlichkeit für aktuelle Studien zur Segmentierung von Konsumenten auf modernen Luxusmärkten wird auch bei Ascheberg, Meurer und Österling (2012) aufgezeigt.

Nachfragetrends auf dem Markt für Lebensmittel

Ein Blick auf das Ernährungsverhalten in der BRD lässt erkennen, dass sich dieses in den letzten Jahren aufgrund von zahlreichen gesellschaftlichen Veränderungen und Marktentwicklungen gewandelt hat (Kirig & Rützler, 2007; Lüth, 2005). Laut Grunert (2006) hat sich die Verbrauchernachfrage immer dynamischer, komplexer und differenzierter entwickelt. Im Hinblick auf den Informationsstand der Verbraucher betrachten Böhm et al. (2007) vier qualitätsbildende Eigenschaften im Nahrungsmittelbereich: Such-, Erfahrungs-, Vertrauens- und Potemkineigenschaften. Wobei die so genannten Sucheigenschaften direkt beim Kauf, die Erfahrungseigenschaften erst beim Gebrauch bzw. Verbrauch, d.h. nach dem Kauf, die Vertrauenseigenschaften nur durch Drittinstitutionen und die Potemkineigenschaften am Produkt nicht mehr direkt nachgeprüft werden können.

Die Ergebnisse der 2012 veröffentlichten Nestlé-Studie (Nestlé Deutschland AG, 2012) zeigen, dass die Qualität der Lebensmittel als ein dominantes Einkaufskriterium der Verbraucher ist. Lebensmittelqualität sei hier definiert als Summe aller Merkmale und bewertbaren Eigenschaften, die nach den subjektiven Kategorien Genusswert, Eignungswert und Gesundheitswert unterteilt wird (Leitzmann, 1993). Im Rahmen der Nestlé-Studie (Nestlé Deutschland

AG, 2012) wurden insgesamt vier Dimensionen von Lebensmittelqualität näher kategorisiert: Genuss/Geschmack, Sicherheit/Transparenz, Gesundheit und Nachhaltigkeit. Laut Nitzko und Spiller (2014) sind auf dem Lebensmittelmarkt Produkte wie Delikatessen und Feinkost verfügbar, um das Bedürfnis nach Genuss zu stillen. Diese füreinander stehenden Synonyme können als Speisen oder Getränke bezeichnet werden, die durch einen besonderen Wohlgeschmack gekennzeichnet sind und bei denen es sich häufig um sehr luxuriöse oder kostspielige Lebensmittel handelt (Zühlsdorf & Spiller, 2012). Des Weiteren gilt der Geschmack als ein hedonistisches Motiv, der als ein Hauptkriterium für die Bewertung von Lebensmitteln gesehen werden kann (Brunsø, Fjord, & Grunert, 2002). Nach Zühlsdorf und Spiller (2012) sind die Lust am besonderen Geschmack und das Wissen um erstklassige Produktqualitäten, auch „sophisticated consumption" genannt, für eine zunehmende Zahl von Konsumenten von großer Wichtigkeit. Die Dimension der Sicherheit und Transparenz hat aus Verbrauchersicht ebenfalls an Bedeutung gewonnen. Durch immer häufiger auftretende Skandale innerhalb der Land- und Ernährungswirtschaft verlangen verunsicherte Konsumenten zum Beispiel mehr Transparenz in Form von Herkunftsnachweise auf den Produkten. Aus diesem Grund verstärken immer mehr Akteure der Land- und Ernährungswirtschaft den Einsatz dieser Nachweise, um die Rückverfolgbarkeit ihrer Produkte bis auf das Feld beziehungsweise den Stall zu gewährleisten (SGS, 2014).

Der Gesundheitstrend geht mit einer bewussteren Ernährung einher. Laut Nitzko und Spiller (2014) kommt dem Wunsch nach physischer und psychischer Gesundheit ein großer Stellenwert zu. Ein Trend, der auch gesamtgesellschaftlich erkennbar ist, denn das Bewusstsein für die Bedeutung der Ernährung im Zusammenhang mit ihren möglichen Folgen für die Gesundheit ist in den letzten Jahren angestiegen (Zühlsdorf & Spiller, 2012). Der Lebensmittelmarkt reagiert hier beispielsweise mit funktionellen Lebensmitteln, auch als „functional food" bezeichnet, die in ihrer Beschaffenheit auf diese Thematik zugeschnitten sind. Laut Zühlsdorf und Spiller (2012) handelt es sich hierbei um modifizierte Produkte, die einen positiven Effekt auf die Gesundheit der Verbraucher bzw. auf ihr Wohlbefinden haben sollen, beispielsweise durch Zusatz von Vitaminen oder Mineralstoffen. Ein steigendes Bewusstsein im Konsumverhalten spiegelt auch die vierte Dimension der Lebensmittelqualität wider: Nachhaltigkeit ist laut GFK (2013) keine kurzfristige Marktentwicklung, sondern ein bestehender Trend, bei dem Lebensmittel und Ernährung eine zentrale Rolle spielen. Aus dieser Tendenz heraus hat sich der Lifestyle of Health and Sustainability (LOHAS) entwickelt. Es handelt sich hierbei um das Bestreben, durch bewussteren Konsum den individuellen Lebensstil, Gesundheit und Genuss mit der gesellschaftlichen Aufgabe nachhaltigen Wirtschaftens in Einklang zu bringen (GFK, 2013). Nach Prüne (2012) setzen sich LOHAS aktiv mit ihrem soziokulturellen Umfeld auseinander, anstatt sich ihm passiv auszusetzen. Hinsichtlich der Konsumentensegmentierung gehört mittlerweile jeder siebte Verbraucher nach seinen Einstellungen sowie seinem Einkaufsverhalten zur Kerngruppe. Somit ist diese Gruppierung seit dem Jahr 2007 um fast fünfzig Prozent gewachsen (GFK, 2013).

Des Weiteren formulieren Kirig und Rützler (2007) in ihrer Studie am Zukunftsinstitut, dass sich an insgesamt acht ermittelten Entwicklungen bzw. *Megatrends,* auf dem Lebensmittelmarkt der gesamtgesellschaftliche Wandel erkennen lässt. Diese lassen sich durch die Schlagworte Gesundheit, Globalisierung, Neo-Ökologie, New Work, Individualisierung, Design und Genuss zusammenfassen. Es heißt, dass sie die Essgewohnheiten der Konsumenten massiv beeinflussen. Kirig und Rützler (2007) bezeichnen Lebensmittel als den Luxusmarkt der Zukunft. Auch hierbei seien Parameter wie Nachhaltigkeit von großer Wichtigkeit, denn Produkte, die mit ethisch-ökologischen Bedenken behaftet sind, werden weniger nachgefragt. Der moderne Konsument hegt dabei eine Auffassung von Luxus, bei der es nicht um Prestige aufgrund von Konsum geht, sondern um die Generierung eines so genannten epikurischen Mehrwerts.

Laut Felder (1997) bedeutet Luxus somit heutzutage nicht mehr, zu beeindrucken, zu protzen oder zur Schau zu stellen. An die Stelle des klassischen Statuskonsums treten Parameter, die beispielsweise zur Steigerung des persönlichen Wohlbefindens dienen.

Diese Entwicklungen auf dem Lebensmittelmarkt zeigen, dass das Entscheidungsverhalten der Konsumenten hinsichtlich ihrer Ernährung zunehmend differenzierter und mit einem hohen Involvement verbunden ist. Laut Kröber-Riel und Gröppel-Klein (2013) liegt ein hohes Involvement vor, wenn der Konsument bereit ist, sich zu engagieren und sich intensiv mit einer Entscheidung auseinander zu setzen. Demzufolge ist dies mit starken emotionalen und kognitiven Prozessen verbunden. Das Involvement ist zunächst eine individuell unterschiedlich ausgeprägte Variable, denn für Verbraucher, die sich nicht detailliert mit ihrer Ernährung beschäftigen, stellen Lebensmittel Low-Involvement-Produkte dar (Spiller, 2010). Andererseits zeigen aktuelle Trends auf dem Lebensmittelmarkt, dass sich Konsumenten zunehmend mit Entscheidungen ihrer Ernährung auseinander setzen und dass somit ein hohes Ernährungsinvolvement vorhanden ist. Laut Spiller (2010) ist beispielsweise ein Gourmet vermehrt auf der Suche nach neuen Feinschmeckerprodukten und beschäftigt sich dementsprechend eingehend mit dieser Thematik. Hierbei zeigen sich Formen des *sophisticated consumption,* der Produktkompetenzen sowie das Wissen um besondere Anbau- und Herstellungsweisen zum Ausdruck bringt (Nitzko & Spiller, 2014).

Nitzko & Spiller (2014) belegen, dass der hohe Differenzierungsgrad der Lebensmittelbranche eine Konzentration der Marketingaktivitäten auf einzelne Marktbereiche verlangt. Diese differenzierten Ansprüche der Konsumenten können in unterschiedliche Themenkomplexe unterteilt werden, um letztendlich die Marketingaktivitäten innerhalb der Ernährungswirtschaft an die jeweilige Zielgruppe anzupassen.

Fragestellung

Unter dem Begriff Konsumentenverhalten versteht man das beobachtbare „äußere" und das nicht beobachtbare „innere" Verhalten von Personen beim Kauf und Konsum wirtschaftlicher Güter. In diesem Kontext hat die Konsumentenverhaltensforschung das Bestreben, die Fragen nach dem Grund des Käuferverhaltens und dessen Ausprägungen zu beantworten (Kröber-Riel und Gröppel-Klein, 2013). Laut Padilla Bravo (2013) lässt sich das Konsumentenverhalten auf dem Lebensmittelmarkt schwer vorhersagen und stellt folglich eine komplexe Aufgabe dar. Wie im vorhergehenden Kapitel thematisiert, entwickelt sich die Nachfrage auf dem Markt für Lebensmittel immer differenzierter, da sich immer mehr unterschiedliche Ernährungsstile etablieren. Als ein existierender Trend ist der Konsum von Delikatessen/Feinkost zu nennen. Nitzko und Spiller (2014) zeigen, dass der Genuss, der beispielsweise durch den Konsum von Delikatessen oder Feinkost befriedigt wird, beim Essen eine zunehmend wichtige Rolle spielt. Damit korrespondierend gewinnt die Feinkostbranche auf dem Lebensmittelmarkt an Relevanz: Laut Nielsen (2014) beläuft sich der Umsatz mit Feinkost im Lebensmitteleinzelhandel im ersten Halbjahr 2014 auf 1,259 Millionen Euro. Des Weiteren belegt das Institut für Demoskopie Allensbach (IFD) (2014), dass in der BRD rund 8,22 Millionen Personen in den zum Befragungszeitpunkt letzten 14 Tagen Feinkostspezialitäten gekauft haben. Insbesondere einkommensstarke Verbraucher mit einem hohen Produktinvolvement stehen im Fokus von Premium- und Gourmetproduzenten. Dieses Kundensegment legt verstärkt Wert auf Produktkompetenzen sowie besondere Anbau- und Herstellungsweisen (Nitzko & Spiller, 2014).

Bis dato liegen allerdings keine empirischen Untersuchungen auf dem Markt für Feinkost/Delikatessen vor, die die für die Konsumenten in der BRD relevanten Faktoren beim Kauf von hochpreisigen Lebensmitteln näher beleuchten.

Es existieren jedoch wissenschaftliche Papiere hinsichtlich der für die Konsumenten relevanten Faktoren auf dem allgemeinen Luxusgütermarkt (z.B. Dubois, Laurent & Czellar, 2001;

Wiedmann, Hennigs & Siebels, 2007, 2009, Hanzaee, Teimourpour & Teimourpour, 2013). Vor diesem Hintergrund soll die vorliegende Arbeit einen Beitrag zur Erforschung des Konsumentenverhaltens innerhalb der Zielgruppe für hochpreisige Lebensmittel leisten. Es wird die Forschungsfrage beantwortet, welche Produktdimensionen Konsumenten beim Kauf von Feinkost/Delikatessen leiten und in welchem Zusammenhang diese mit dem allgemeinen Lebensmittelmarkt stehen.

Methodik

Als Umfragemedium dient ein standardisierter Online-Fragebogen, in welchem die Teilnehmer zu ihrer Einstellung im Hinblick auf Luxuslebensmittel befragt werden. Die Bearbeitung des Fragebogens nahm etwa 10 Minuten in Anspruch.

Der Fragebogen war für die Teilnehmer als Online-Umfrage vom 30. Januar bis 17. Februar 2014 zugänglich. Die Akquise erfolgte über verschiedene Kanäle, hauptsächlich jedoch über eine E-Mail-Ansprache[14], über Flyer mit einem Hinweis auf die Umfrage, ausgelegt in ausgewählten Feinkostgeschäften in Südniedersachsen und Nordhessen, sowie mithilfe des sozialen Netzwerks „Facebook". Hierbei ist zu erwähnen, dass nur Personen an der Befragung teilnehmen konnten, die zum Zeitpunkt der Umfrage mindestens 25 Jahre alt gewesen sind. Damit sollte sichergestellt werden, dass die Probanden regelmäßig aktive Kaufentscheidungen auf dem Lebensmittelmarkt treffen.

Die Konzeption des Fragebogens ist angelehnt an Wiedmann, Hennigs und Siebels (2009) und wurde in Bezug auf die Thematik Lebensmittel entsprechend erweitert. Der Fragebogen gliedert sich inhaltlich sowie formal in vier Abschnitte. Nach einer kurzen Einleitung werden die Teilnehmer gebeten, Fragen hinsichtlich ihrer Ernährung zu beantworten. Im zweiten Abschnitt wird das Lebensmittel-Kaufverhalten der Teilnehmer näher erfragt. Hier wird nach Lebensmitteln im Allgemeinen und nach Feinkost/ausgewählten Delikatessen unterteilt. Der dritte Abschnitt des Fragebogens beschäftigt sich mit der Thematik der Luxuslebensmittel. Hier sind die Teilnehmer beispielsweise aufgefordert, 12 Bilder von verschiedenen hochpreisigen Lebensmitteln bezüglich ihrer subjektiven Assoziation von Luxus zu bewerten. Der vierte Fragebogenabschnitt dient hauptsächlich der Erfassung der soziodemografischen Parameter der Befragten. Zugleich ist es den Teilnehmern möglich, abschließend Anregungen und Kritik zu äußern.

---

[14] Die Adressen wurden hauptsächlich den Internetseiten von Unternehmen der Feinkostbranche entnommen.

Der Fragebogen besteht zum größten Teil aus geschlossenen Fragen, die auf einer fünfstufigen Likert-Skala bzw. durch einfache Entscheidungen, wie „entweder-oder" oder durch Mehrfachnennungen, beantwortet werden können. Die Beantwortung sämtlicher Fragen erfolgte anonym.

Die statistische Auswertung der Daten wurde mit dem Programm IBM SPSS Statistics 21 durchgeführt. Nebst deskriptiven Auswertungen von sozio-demografischen Parametern beinhaltet diese eine explorative Faktorenanalyse (Hauptkomponentenanalyse, Varimax Rotationsmethode mit Kaiser-Normalisierung), bei der die Likert-skalierten Aussagen der Fragen – *„Welche Kriterien beeinflussen Sie beim Kauf von Lebensmitteln?"* und *„Was erwarten Sie im Vergleich zu günstigen Lebensmitteln, wenn Sie hochpreisige Lebensmittel einkaufen?"* – anhand multivariater Analysen zu wenigen Faktoren verdichtet wurden. Die beiden Fragen ermöglichen es, den Forschungsgegenstand näher zu betrachten, da sie die beeinflussenden Faktoren sowie die Erwartungshaltung im Kontext auf hochpreisige Lebensmittel thematisieren.

Bei der explorativen Faktorenanalyse handelt es sich um ein iteratives Verfahren zur Dimensionsreduktion. Entsprechend ist es dessen Ziel, eine Vielzahl von Items auf der Grundlage ihrer korrelativen Beziehung zu einer Anzahl überschaubarer und interpretierbarer Faktoren zusammenzufassen (Bühl, 2008).

Deskriptive Statistik

Insgesamt konnten die Fragebögen von 220 Teilnehmern ausgewertet werden. Tabelle 1 veranschaulicht die deskriptive Auswertung einiger Variablen des Datensatzes.

Tabelle 1. Deskriptive Statistik der Befragungsteilnehmer

Teilnehmercharakteristika	
Durchschnittliches Alter (Jahre)	35,3
Geschlecht [%]:	
Anteil weiblicher Teilnehmer	65,5
Anteil männlicher Teilnehmer	34,5
Herkunft [%]:	
Anteil Teilnehmer aus Niedersachsen	38,2
Anteil Teilnehmer aus Hessen	20,5
Anteil Teilnehmer aus Bayern	12,7
Sonstige	28,6
Familienstand [%]:	
ledig	35,5
in einer Beziehung lebend	34,5
verheiratet (mit Kindern)	21,4
Sonstige	8,6
Höchster berufsbildender Abschluss [%]:	
Anteil Teilnehmer (Fach-) Hochschulabschluss	57,5
Sonstige	42,5
Berufliche Ausrichtung [%]:	
Angestellte	35,6
Studierende	28,3
Selbstständige (mit Angestellten)	8,7
Sonstige	27,4
Monatliches Nettohaushaltseinkommen [%]:	
< 2.000 Euro	45,5
2.000 - 3.999 Euro	33,2
≥ 4.000 Euro	14,2
Keine Angabe	7,1

Einkaufs- und Ernährungsverhalten der Teilnehmer	
Haupteinkaufsstätten [%][1]	
Supermarkt (z.B. Edeka, Rewe, Tengelmann, etc.)	83,2
Discounter (z.B. Aldi, Lidl, Netto, Norma, Penny, etc.)	51,4
Bioladen (z.B. Alnatura)	16,4
Feinkosthandel	5,0
Ø monatliche Ausgaben für Lebensmittel pro Person [%]:	
< 250 Euro	35,0
250 bis 499 Euro	47,3
> 500 Euro	15,4
Keine Angabe	2,3
Ernährungstypen [%][1]:	
fleischlastige Ernährung	27,3
fleischarme Ernährung	24,5
Veganer	2,3
Vegetarier	0,9

[1]Mehrfachnennungen waren möglich, nur auffällige Kategorien genannt

Das Durchschnittsalter der Befragten beträgt 35,3 Jahre und liegt somit unter dem arithmetischen Mittel der deutschen Bevölkerung mit 45,5 Jahren bei Frauen und 42,8 Jahren bei Männern (BIB, 2014). Der älteste Befragungsteilnehmer weist ein Alter von 74 Jahren auf.

65,5% der Teilnehmer sind weiblichen Geschlechts. Hinsichtlich ihrer Herkunft geben 38,2% das Bundesland Niedersachsen an, gefolgt von Hessen (20,5%) und Bayern (12,7%). 35,5% der Befragten bezeichnen ihren Familienstand als ledig, 34,5% leben in einer Beziehung und 21,4% sind verheiratet (mit Kindern). Über die Hälfte der Partizipanten besitzen einen (Fach-)Hochschulabschluss. Folglich liegen diese 57,5% mit ihrem höchsten bildenden Abschluss über dem bundesdeutschen Durchschnitt von 27,3% (Statistisches Bundesamt, 2013a). Des Weiteren geben 17,8% eine Lehrausbildung als höchsten berufsbildenden Abschluss an; 9,1% befanden sich zum Zeitpunkt der Befragung noch in der Ausbildung. In Bezug auf die berufliche Ausrichtung arbeiten 35,6% im Angestelltenverhältnis, gefolgt von Studierenden (28,3%) und Selbstständigen mit oder ohne Angestellten (8,7%).

Das monatliche Nettohaushaltseinkommen liegt bei dem größten Teil der Befragten bei unter 2.000 Euro (45,5%), 33,2% verfügen über ein Nettohaushaltseinkommen von 2.000-3.999 Euro und 14,2% über ein Nettohaushaltseinkommen von 4.000 Euro oder mehr monatlich, wobei 7,1% der Teilnehmer keine Angaben zu diesem Thema machen wollten. Damit ist das

Einkommensniveau in dieser Stichprobe insgesamt niedriger als in der gesamtdeutschen Bevölkerung (Institut für Demoskopie Allensbach, 2013).

Des Weiteren werden zur Stichprobenbeschreibung Fragen zum Einkaufs- und Ernährungsverhalten der Partizipanten gestellt. Bei den Haupteinkaufsstätten geben 83,2% den Supermarkt an, über die Hälfte (51,4%) den Discounter, 16,4% den Bioladen sowie 5,0% den Feinkosthandel. In der NVS II wurden ebenfalls Supermärkte und Discounter als primäre Einkaufsstätten ermittelt (Max Rubner-Institut, 2008). Die durchschnittlichen monatlichen Ausgaben für Lebensmittel liegen bei dem Hauptteil der Befragten (47,3%) bei 250 bis 499 Euro, 35,0% verwenden unter 250 Euro und 15,4% über 500 Euro hierfür. 2,3% äußerten sich hierzu nicht. In Bezug auf das Ernährungsverhalten bezeichnen 27,3% diese als fleischlastig sowie 24,5% als fleischarm. Der Anteil der Veganer liegt mit 2,3% höher als der Anteil der Vegetarier mit 0,9%. Der Vegetarieranteil in der Bundesrepublik beträgt 1,6% (Max Rubner-Institut, 2008).

Zusammenfassend kann gesagt werden, dass die Teilnehmer im Vergleich zu repräsentativen Zahlen für Deutschland jung sind, einen überdurchschnittlich hohen Bildungsgrad und ein unterdurchschnittlich hohes Einkommensniveau aufweisen (Institut für Demoskopie Allensbach, 2013; Statistisches Bundesamt, 2013b).

Abbildung 2 zeigt die relativen Häufigkeiten zur Bedeutung von Ernährung, Genuss und Qualität im Alltag der Befragten auf. Die Teilnehmer legen hier einen großen Stellenwert auf ihre Ernährung und den Genuss. Eine höhere Zahlungsbereitschaft für Lebensmittel mit mehr Qualität und Genuss besitzen 65% der Befragten. 37,8% würden hierfür 10-19,99% mehr ausgeben.

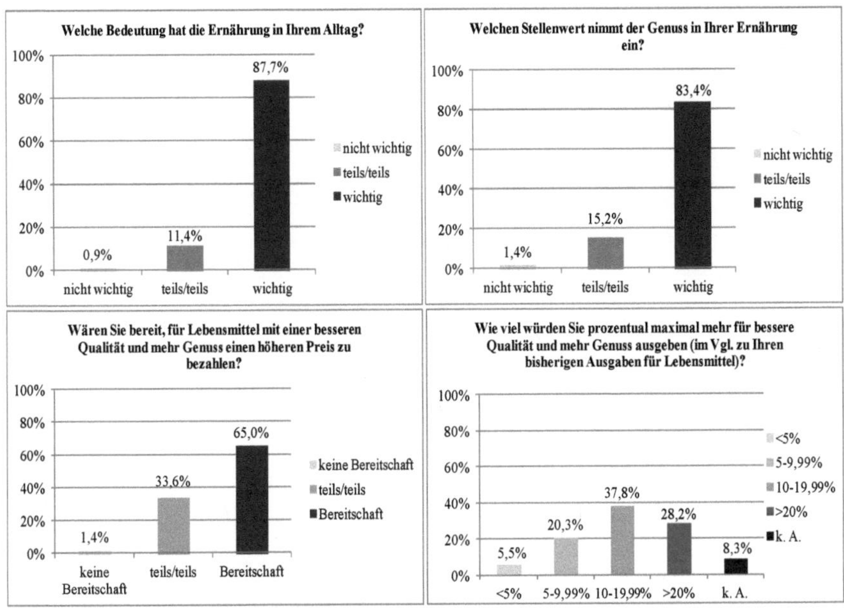

Abbildung 2. Relative Häufigkeiten im Antwortverhalten zur Bedeutung von Ernährung, Genuss und Qualität im Alltag der Befragten[15]

---

[15] Nicht alle Fragen wurden von allen Teilnehmern (N=220) beantwortet. Die Anzahl abgegebener Antworten schwankt zwischen 217 und 220.

Faktorenanalyse

Die Faktorenanalyse hat eine 4-Faktorlösung, bestehend aus insgesamt 16 Variablen, ergeben. Tabelle 2 fasst die vier Einstellungsdimensionen zusammen.

Tabelle 2. Ladungen und Verteilungsparameter der Faktorenanalyse[1] (Hauptkomponentenanalyse) auf Basis von Varimax-Rotationsmethode mit Kaiser-Normalisierung[2]

Faktor 1: Marketing und Prestige (Cronbachs Alpha: 0,701, erklärter Anteil der Gesamtvarianz: 17,94%)	Ladung	μ; σ
Beim Kauf von Lebensmitteln beeinflusst mich …		
… eine erfolgreiche Marke.	0,709	3,11; 0,942
… ein schönes Einkaufsgefühl.	0,706	2,73; 1,023
… eine ansprechende Werbung.	0,698	3,41; 0,942
… eine schöne Ladenatmosphäre.	0,697	2,57; 0,952
… wenn das Produkt optisch ansprechend ist.	0,639	2,20; 0,833
Wenn ich hochpreisige Lebensmittel kaufe, dann erwarte ich…		
…. ein gutes Ansehen bei Familie, Freunden und Bekannten.	0,556	3,29; 1,094
Faktor 2: Vertrauen und Nachhaltigkeit (Cronbachs Alpha: 0,797, erklärter Anteil an der Gesamtvarianz: 17,66%)	Ladung	μ; σ
Wenn ich hochpreisige Lebensmittel kaufe, dann erwarte ich…		
… faire Herstellungsbedingungen für Mensch und Tier.	0,854	2,05; 0,999
… eine bessere Rückverfolgbarkeit/Produkttransparenz.	0,814	2,21; 0,974
… einen vertrauenswürdigen Hersteller.	0,648	1,95; 0,933
… einen höheren Gesundheitswert im Vergleich zu „Discount"-Produkten.	0,643	2,47; 1,087
Beim Kauf von Lebensmitteln beeinflusst mich…		
… eine objektive Zertifizierung/ein vertrauenswürdiges Label (z.B. Bio, Fair Trade, Tierwohl)	0,674	2,72; 1,088
Faktor 3: Preis-Leistungsverhältnis (Cronbachs Alpha: 0,522, erklärter Anteil der Gesamtvarianz: 11,22%)	Ladung	μ ; σ
Beim Kauf von Lebensmitteln beeinflusst mich…		
… wenn das Preis-Leistungsverhältnis stimmt.	0,853	1,90; 0,797
… der Preis.	0,792	2,42; 0,890
… wenn ich bereits gute Erfahrungen mit dem Produkt gemacht habe.	0,449	1,44; 0,558

Faktor 4: Hedonismus und Sensorik (Cronbachs Alpha: 0,747, erklärter Anteil der Gesamtvarianz: 10,81%)	Ladung	μ ; σ
Wenn ich hochpreisige Lebensmittel kaufe, dann erwarte ich…		
… mehr Genuss.	0,860	1,78; 0,774
… einen besseren Geschmack.	0,772	1,65; 0,750

[1]Aussagen waren anhand von fünfstufiger Likert-Skala zu beantworten: von „stimme voll zu" (1) bis „stimme ganz und gar nicht zu" (5)

[2]Kaiser-Meyer-Olkin-Kriterium: 0,748, höchste Signifikanz der Werte nach Bartlett (p=,000), erklärter Anteil der Gesamtvarianz basierend auf der rotierten Summe der quadrierten Ladungen: 57,62%

Als Gütekriterien für Faktorenanalysen gelten unter anderem das Kaiser-Meyer-Olkin-Kriterium (KMK), der Bartlett-Test auf Sphärizität und der erklärte Anteil der Gesamtvarianz. Zunächst wird das KMK näher betrachtet. Laut Backhaus et al. (2006) zeigt dieses an, in welchem Umfang die Ausgangsvariablen zusammengehören, es dient als Maß für die Stichprobeneignung und ist somit ein geeigneter Indikator, ob die Faktorenanalyse durchgeführt werden sollte. Ein KMK kleiner als 0,5 eignet sich gemäß der Literatur nicht für eine Faktorenanalyse, somit liegt der Wert in der durchgeführten Faktorenanalyse mit 0,748 in dem geforderten Wertebereich und kann als „ziemlich gut" beschrieben werden (BACKHAUS et al. 2006: 276).

Des Weiteren stellt sich der Bartlett-Test (Test auf Spherizität) als Maß für die Angemessenheit der Stichprobe dar. Er überprüft die Nullhypothese H_0, dass die Stichprobe aus einer Grundgesamtheit entstammt, in der die Variablen unkorreliert sind (Backhaus et al., 2006). In der vorliegenden Faktorenanalyse erreicht der Bartlett-Test einen höchst signifikanten Wert, sodass die Nullhypothese dieses Tests mit einer Irrtumswahrscheinlichkeit von p=0,000 abgelehnt werden kann. Folglich kann man davon ausgehen, dass in der Grundgesamtheit korrelative Zusammenhänge zwischen den insgesamt 16 Variablen bestehen.

Die erklärte Gesamtvarianz wird ebenfalls als Kriterium für die Güte der Faktorenanalyse angesehen. In diesem Fall erklären die vier ermittelten Faktoren 57,62% der Ausgangsvarianz.

Als weiteres Gütekriterium bewertet der Cronbachs Alpha-Wert die interne Konsistenz zwischen den Faktoren und dient somit zur Konsistenzüberprüfung. Hier wird in der Literatur ein Wert von 0,7 oder besser verlangt (Panayides nach Nunnally, 2009). Drei von vier Faktoren weisen mit Werten um 0,7 einen guten Wert für Cronbachs Alpha auf. Lediglich Faktor 3

Preis- Leistungsverhältnis liegt mit einem Wert von 0,522 unter dem in der Literatur angestrebten Wert. Allerdings verbessert sich das Cronbachs Alpha nur leicht in der zweiten Nachkommastelle, wenn man das am wenigsten trennscharfe Item – „wenn ich bereits gute Erfahrungen mit dem Produkt gemacht habe" – aus der Skala entfernt. Zusammenfassend kann bei drei von vier Faktoren eine gute Konsistenz bestätigt werden.

Die aus der Hauptkomponentenanalyse resultierenden vier Faktoren spiegeln die Kaufkriterien von Lebensmitteln und die Erwartungshaltungen beim Kauf von hochpreisigen Lebensmitteln seitens der Befragten wider. Faktor 1 *Marketing und Prestige* beinhaltet sechs Variablen, wobei sich die ersten fünf auf den Aspekt Marketing beziehen, die sechste Variable betrifft die Thematik des demonstrativen Konsums. Die Mittelwerte des Antwortverhaltens der Teilnehmer zeigen bei diesen sechs Items eine vermehrt neutrale Ausprägung, wenn man von der verwendeten fünfstufigen Likert-Skala ausgeht, bei der die Antwortmöglichkeit „teils/teils" den Wert 3 erhält. Die Variable „Wenn ich hochpreisige Lebensmittel kaufe, dann erwarte ich ein gutes Ansehen bei Familie, Freunden und Bekannten" weist mit einem Wert von 1,094 die höchste Standardabweichung in der 4-Faktorlösung auf und folglich die höchste Heterogenität im Antwortverhalten der Partizipanten. Mit 17,94% hat dieser Faktor den höchsten erklärten Anteil an der Gesamtvarianz.

Faktor 2 *Vertrauen und Nachhaltigkeit* beinhaltet fünf Variablen, wobei sich die ersten vier Variablen auf hochpreisige Lebensmittel beziehen. Hier ist der höchste Mittelwert und somit eine positive Zustimmung bei dem Item „Wenn ich hochpreisige Lebensmittel kaufe, dann erwarte ich einen vertrauensvollen Hersteller" (μ=1,95) zu verzeichnen.

Mit der Variable „Beim Kauf von Lebensmitteln beeinflusst mich, wenn ich bereits gute Erfahrungen mit dem Produkt gemacht habe" beinhaltet Faktor 3 *Preis- Leistungsverhältnis* die Variable, die in der 4-Faktorlösung den höchsten Mittelwert (μ=1,44) und die geringste Standardabweichung (σ=0,558) vorweisen kann. Daher stimmen die Befragten tendenziell zu, dass gute Erfahrungen mit dem Produkt sie beim Kauf stark beeinflussen. Das Antwortverhalten ist in Bezug auf dieses Item am homogensten.

Des Weiteren beinhaltet Faktor 4 *Hedonismus und Sensorik* zwei Variablen. Die Befragten erwarten hier beim Kauf von hochpreisigen Lebensmitteln mehr Genuss sowie einen besseren Geschmack. Beide Items haben, neben der bereits angesprochenen Variable in Faktor 3, die höchste Zustimmung im Antwortverhalten (μ=1,65) und die geringste Standardabweichung (σ=0,750). Ferner erklärt der letzte Faktor mit 10,81% den geringsten Anteil an der Gesamtvarianz

Jedoch ist bei der vorliegenden Stichprobe mit einer Teilnehmerzahl von 220 Personen keine Repräsentativität gegeben, da nicht auf die Grundgesamtheit der bundesdeutschen Bevölkerung zu schließen ist.

Diskussion

Ziel der vorliegenden Arbeit ist es zu analysieren, welche Variablen und nutzenbasierten Dimensionen Konsumenten beim Kauf von Produkten auf dem Markt für hochpreisige Lebensmittel beeinflussen und welche Gemeinsamkeiten mit dem Konsumentenverhalten auf dem allgemeinen Lebensmittelmarkt daraus abgeleitet werden können.

Auf dem Markt für allgemeine Luxusgüter wurden nach Wiedmann, Hennigs, und Siebels (2007, 2009) vier nutzenbasierte Dimensionen (finanzieller, funktioneller, individueller und sozialer Wert) präzisiert, die sich ebenfalls in den Ergebnissen der eigenen Faktorenlösung wieder finden. Demzufolge ist es ersichtlich, dass große Überschneidungen zwischen den Ergebnissen für den allgemeinen Luxusmarkt und den Markt für hochpreisige Lebensmittel bestehen.

In Faktor 1 zeigt sich, dass im Hinblick auf hochpreisige Lebensmittel die Variable „(…) ein gutes Ansehen bei Familien, Freunden und Bekannten" Bedeutung hat. Ein gutes Ansehen bezieht sich nach Wiedmann, Hennigs, und Siebels (2007, 2009) auf den sozialen Wert des Luxuskonsums und damit auf die Signalwirkung von Luxus innerhalb einer sozialen Gruppe. Prestige und der demonstrative Konsum zur Geltung in Sozialgefügen gehören zu den traditionellen Motiven für Luxuskonsum (Veblen, 1899). Da es sich bei Faktor 1 um die Variable mit dem heterogensten Antwortverhalten handelt, ist abzuleiten, dass sich die Einstellungen von Konsumenten zur sozialen (traditionellen) Komponente des Luxuskonsums als sehr gemischt darstellen. Denn laut Nitzko und Spiller (2014) entwickeln sich verstärkt Formen des *sophisticated consumption* und diese Konsumausprägung tritt vermehrt an die Stelle des klassischen Statuskonsums. Yeoman und McMahon-Beattie (2006) zeigen in diesem Kontext auf, dass sich innerhalb des Luxusverständnisses ein Wandel vollzogen hat, bei dem es sich nicht mehr um Status, sondern vielmehr um Erfahrungen und Genuss handelt. Zu den Erfahrungseigenschaften zählt beispielsweise die Sensorik. Jedoch können diese erst nach dem Gebrauch bzw. Verbrauch eines Produktes seitens der Konsumenten bestimmt werden, somit beeinflussen diese erst zukünftige Kaufentscheidungen (Böhm, 2007).

Der zweite Faktor beinhaltet vier Items zu Nachhaltigkeit und zeigt somit, dass der Thematik auch in dieser Stichprobe ein großer Stellenwert zukommt. Mit dem Aspekt der Nachhaltigkeit liegt hier eine Erweiterung des Modells für Dimensionen von Luxus im Allgemeinen von Wiedmann, Hennigs und Siebels (2009) vor. Laut Yeoman und McMahon-Beattie (2006) erscheint es von großer Wichtigkeit, dass die durch den Wandel im Luxuskonsum hinzukommenden Wertehaltungen der Konsumenten betrachtet werden, um die Schlüsselfaktoren aus Konsumentensicht zu identifizieren und absatzpolitische Maßnahmen einzuleiten und anzupassen. Die Konsequenz des veränderten Konsumverhaltens auf Luxusmärkten sei, dass der heutige Luxuskonsument einerseits eine höhere Erwartungshaltung an die Produkte aufweist, aber andererseits auch bereit ist, mehr finanzielle Mittel für eine Steigerung des Wohlbefindens aufzuwenden.

Des Weiteren bezieht sich Faktor 3 auf den allgemeinen Lebensmittelmarkt. Hier lassen sich die Konsumenten durch das Preis-Leistungsverhältnis beeinflussen. Wiederum liegt eine Übereinstimmung mit Wiedmann, Hennigs und Siebels (2009) vor, da es sich hierbei einmal um den finanziellen Wert, aber auch um den funktionalen Wert handelt. Dieser Zusammenhang zeigt sich in der Verbraucherentscheidung, indem Verbraucher den Preis auch als Indikator für Qualität nutzen. Damit wird ihnen eine Möglichkeit geboten, ihr Risiko für Fehlkäufe zu reduzieren (Böhm, 2007).

Faktor 4 beinhaltet Variablen, die sich auf Hedonismus und Sensorik bei hochpreisigen Lebensmitteln beziehen. Hier findet sich der individuelle Wert nach Wiedmann, Hennigs und Siebels (2009) wieder. Kisabaka (2001) thematisiert die Affinität von Luxuskonsumenten zu Hedonismus, der unter anderem das Bedürfnis zur Selbstentfaltung als maßgebliches Motiv beim Luxuskonsum inkludiert.

Zusammenfassend lässt sich sagen, dass bei den Faktoren 1 und 2 sowohl der Markt für hochpreisige Lebensmittel als auch der allgemeine Lebensmittelmarkt thematisiert werden. Faktor 3 bezieht sich indes nur auf den allgemeinen Lebensmittelmarkt und Faktor 4 auf den Markt für hochpreisige Lebensmittel. Dementsprechend wird gezeigt, dass die Dimensionen *Marketing und Prestige* sowie *Vertrauen und Nachhaltigkeit* auf beiden Märkten relevant sind, während das Preis-Leistungsverhältnis eher auf dem allgemeinen Lebensmittelmarkt bzw. *Hedonismus und Sensorik* eher auf dem Markt für hochpreisige Lebensmittel zum Tragen kommen.

Nach Liebel (2007) geht die Means-End-Methode davon aus, dass Menschen in ihrem Leben bestimmte Wertvorstellungen und Zielsetzungen haben, die sich auf ihr Konsumverhalten auswirken, indem sie einen Zusammenhang zwischen den Eigenschaften eines Produktes, den sich ergebenden individuellen Konsequenzen in ihrem Nutzen und ihrer Wertehaltung herstellen. Folglich wird das Produkt zum Mittel (means), um ein bestimmtes Ziel (ends) zu erreichen. Die Theorie findet in der Kognitionspsychologie Anwendung, denn sie basiert auf der Vorstellung, dass der Verbraucher seine Bedürfnisse selbst verstehen und verbalisieren kann (Spiller, 2010).

In Anlehnung an die Methode können diese Zusammenhänge auch bei den ermittelten Faktoren dargestellt werden, die in Tabelle 3 exemplarisch aufgezeigt werden.

Tabelle 3. Schematische Darstellung nach Means-End-Methode nach Gutman und Reynolds (1979)[16]

Faktor	Eigenschaften	Nutzenkomponenten	Wertehaltungen
1) Marketing und Prestige	… wenn das Produkt optisch ansprechend ist.	Man gilt als ästhetischer Mensch, Ansehen (sozial)	Ästhetik
2) Lebensmittelsicherheit und Nachhaltigkeit	… faire Herstellungsbedingungen für Mensch und Tier.	umweltschützend (funktional) <hr> man gilt als umweltbewusst (sozial)	Altruismus
3) Preis-Leistungsverhältnis	… wenn ich bereits gute Erfahrungen mit dem Produkt gemacht habe.	Qualität (funktional)	Vertrauen
4) Hedonismus und Sensorik	… mehr Geschmack.	Sensorik (individuell)	Hedonismus

Im Hinblick auf den ersten Faktor kann in diesem Kontext interpretiert werden, dass Verbraucher beim Kauf von Lebensmitteln einen Fokus auf die Eigenschaft der Optik legen, um folglich als Konsequenz den sozialen Nutzen als ästhetischer Mensch zu gelten zu erreichen. Demzufolge ergibt sich die Wertehaltung der Ästhetik. Laut Kisabaka (2001) stechen Produkte mit einer stark ausgeprägten Ästhetik aufgrund ihres Designs oder Stils hervor.

Der zweite exemplarisch dargestellte Faktor bezieht sich auf hochpreisige Lebensmittel. Hier dienen faire Herstellungsbedingungen für Mensch und Tier als Eigenschaft. Bezug nehmend auf den Nutzen kann dieser einmal in funktionaler und sozialer Weise interpretiert werden. Letztendlich wird die Wertehaltung Altruismus widergespiegelt. Es zeigt sich, dass das Kon-

---

[16] In Anlehnung an Zeithaml (1988)

sumverhalten in der BRD zunehmend geprägt ist durch eine artgerechte Tierhaltung, einen ökologisch-nachhaltigen Anbau und fairen Handel (SGS, 2014).

Des Weiteren thematisiert der dritte Faktor die Eigenschaft, dass bereits gute Erfahrungen mit dem Produkt gemacht wurden. Als Konsequenz zeigt sich der funktionale Nutzen der Qualität. Man beruft sich auf Erfahrungswerte und zeigt damit, dass der Wertehaltung Vertrauen (in die Qualität des Produkts) ein hoher Stellenwert zugewiesen wird. Nach Böhm et al. (2007) können die Erfahrungseigenschaften erst beim Gebrauch bzw. Verbrauch des Produktes, sprich nach dem Kauf, beurteilt werden und stehen erst bei der nächsten Kaufentscheidung zur Verfügung.

Das Konsumentenverhalten in Bezug auf den vierten Faktor ist geprägt durch die Eigenschaft des guten Geschmacks, der als ein hedonistisches Motiv für die Bewertung von Lebensmitteln gesehen werden kann (Brunsø, Fjord & Grunert, 2002). Der individuelle Nutzen liegt hier in der Sensorik, die letztendlich zur Wertehaltung des Hedonismus führt.

Wie mehrfach beschrieben, lässt sich auch an dieser Stelle resümierend erkennen, dass das Ernährungsinvolvement auf den Märkten für allgemeine Lebensmittel und für hochpreisige Lebensmittel steigt. Die Konsumenten in der BRD setzen sich mehr und mehr mit der individuellen Bedeutung und den Folgen ihrer Ernährung auseinander und folglich resultiert daraus ein Wertewandel, der zukünftig in der Wissenschaft sowie in der Land- und Ernährungswirtschaft Beachtung finden muss.

Fazit und Ausblick

In vorausgegangenen Studien, die den Luxuskonsum thematisieren, wurden relevante Einflussfaktoren und dementsprechend nutzenbasierte Dimensionen ermittelt, um die Zusammensetzung des individuellen Luxuswerts näher zu beleuchten. Diese Forschungsarbeiten beschäftigen sich hauptsächlich mit Luxusgütern im Allgemeinen. Jedoch kann im Zuge des Wandels im Luxuskonsum die Etablierung neuer, differenzierter Luxusmärkte beobachtet werden (Silverstein, Fiske & Butman, 2008). Aktuelle Zahlen belegen zum Beispiel, dass die Feinkostbranche wächst und sich hier ein neuer Luxusmarkt abzeichnet. Bisher existiert jedoch keine allgemeingültige Definition für Produkte in diesem Lebensmittelsektor. Um die aus Konsumentensicht wahrgenommene Definition von Luxus-Lebensmitteln zu präzisieren, sollte das Konsumentenverhalten auf den entsprechenden Märkten und die damit verbundenen beeinflussenden Faktoren erforscht werden. Darin besteht die Relevanz der vorliegenden Arbeit, die die Verbraucheransprüche und Produktdimensionen in Bezug auf hochpreisige Lebensmittel ermittelt.

Die Ergebnisse der Auswertung haben gezeigt, dass Überschneidungen zwischen dem in der Literatur angesprochenen allgemeinen Markt für Luxusgüter und dem Markt für hochpreisige Lebensmittel existieren. Es kann festgestellt werden, dass im Ernährungsverhalten und dem Lebensmittelkonsum in Deutschland ein Wertewandel stattfindet.

Der Konsum von hochpreisigen Lebensmitteln wird vornehmlich durch Qualität, Hedonismus und Nachhaltigkeit motiviert, während das Prestige-Motiv zunehmend an Bedeutung verliert. Eine ähnliche Entwicklung wird auch für den allgemeinen Luxusmarkt bestätigt (Meurer & Manniger, 2012; Yeoman, 2011). Insbesondere die Thematik der Nachhaltigkeit wird zunehmend auf beiden Luxusmärkten fokussiert und nimmt bei Konsumenten einen großen Stellenwert ein (Kapferer, 2010). Laut GFK (2013) ist Nachhaltigkeit ein Trend, bei dem Lebensmittel und Ernährung im Allgemeinen zunehmend eine Rolle spielen.

Nitzko und Spiller (2014) thematisieren, dass der hohe Differenzierungsgrad der Lebensmittelbranche eine Konzentration der Marketingaktivitäten auf einzelne Marktbereiche verlangt. Diese differenzierten Ansprüche der Konsumenten finden sich ebenfalls auf dem Markt für hochpreisige Lebensmittel wieder. Eine Betrachtung der Einstellungen und Ansprüche von Verbrauchern dient dazu, um letztendlich die Marketingaktivitäten innerhalb der Ernährungswirtschaft zielgruppenspezifisch anzupassen.

In diesem Zusammenhang zeigen sich Limitierungen der vorliegenden Arbeit. Sie gibt zunächst nur einen Überblick über luxusbestimmende Faktoren, die Konsumenten auf dem Markt für hochpreisige Lebensmittel beeinflussen. Um eine genauere Zielgruppensegmentierung für gezielte Marketingvorhaben zu geben, sei der weiterführenden Forschung empfohlen, die Zielgruppen beispielsweise anhand einer Clusteranalyse näher zu untersuchen.

Literatur

Ascheberg, C., Meurer, J., & Österling, A. (2012). The Luxury Universe–Angebots-und Kundensegmentierung globaler Luxusmärkte als Basis für erfolgreiche Positionierungsstrategien. In C. Burmann, V. König, & J. Meurer (Hrsg.), *Identitätsbasierte Luxusmarkenführung* (S. 85-101). Springer Fachmedien Wiesbaden.

Backhaus, K., Erichson, B., Plinke, W., & Weiber, R. (2006). *Multivariate analysemethoden.* Springer, Berlin.

Böhm, J, De Witte, T., Schulze, H, & Spiller, A. (2007). Preis-Qualitäts-Relationen im Lebensmittelmarkt: Eine Analyse auf Basis der Testergebnisse der Stiftung Warentest, *Diskussionsbeitrag 0702*, Georg-August-Universität Göttingen.

Brunsø, k., Fjord, T. A., & Grunert, K. G. (2002). Consumers´ Food Choice and Quality Perception, *working paper no 77*, The Aarhus School of Business.

Bühl, A. (2008). SPSS 16 Einführung in die moderne Datenanalyse. 11. aktualisierte Auflage, Pearson Studium, München.

Bundesinstitut für Bevölkerungsforschung (BIB) (2014). Durchschnittsalter der Bevölkerung in Deutschland im Jahr 2012. URL:
http://www.bib-demografie.de/ SharedDocs/Glossareintraege/DE/D/durchschnittsalter
_bevoelkerung.html. Zugriff am 23. Oktober 2014.

D´arpizio, C. (2014). *Global Luxury Goods Worldwide Market Study Spring 2014*. Bain and Company (Ed.), Boston.

D´arpizio, C. & Levato, F. (2014). *Lens on worldwide luxury consumer: Relevant segments, behaviors and consumption patterns, nationalities and generations compared*. Bain and Company (Ed.), Boston.

Dubois, B., Laurent, G., & Czellar, S. (2001). Consumer rapport to luxury: Analyzing complex and ambivalent attitudes. *Le Cahiers de Recherche, 736*, HEC Paris.

Felder, A. (1997). *Erfolgsfaktoren im Luxusgütermarketing*. Diplomarbeit, Johannes Gutenberg-Universität Mainz.

GFK Consumer Panels und Bundesvereinigung der Deutschen Ernährungsindustrie (2013) (Hrsg.). Consumers´ Choice ´13: Bewusster Genuss-Nachhaltige Gewinne für Ernährungsindustrie und Konsumenten. 5. Ausgabe.

Grugel-Panier, D. (1996). *Luxus: eine begriffs- und ideengeschichtliche Untersuchung unter besonderer Berücksichtigung von Bernard Mandeville*. Peter Lang Verlag, Frankfurt a.M..

Grunert, K. G. (2006). How changes in consumer behaviour and retailing affect competence requirements for food producers and processors. *Economia Agraria y Recursos Naturales, 6*(11), 3-22.

Hanzaee, K. H., Teimourpour, B., & Teimourpour, B. (2012). Segmenting Consumers Based on Luxury Value Perceptions. *Middle-East Journal of Scientific Research, 12*(11), 1445-1453.

Hellhammer, S. (2007). Marketingforschung im Luxussegment: Mechanismen, Märkte und Methoden, Dissertation, Otto-von-Guericke- Universität Magdeburg.

Institut für Demoskopie Allensbach (2013). AWA Allensbacher Markt- und Werbeträgeranalyse 2013, Allensbach.

Institut für Demoskopie Allensbach (2014). Anzahl der Personen in Deutschland, die innerhalb der letzten 14 Tagen Feinkostspezialitäten gekauft haben, von 2012 bis 2014 (in Millionen). URL: http://de.statista.com/statistik/daten/studie/173667/umfrage/lebensmittel---konsum-von-feinkostspezialitaeten/. Zugriff am 20. Oktober 2014.

Kapferer, J. N. (2010). All that glitters is not green: The challenge of sustainable luxury. *European Business Review*, 40-45.

Kirig, A., & Rützler, M. H. (2007). Food-Styles. *Die wichtigsten Thesen, Trends und Typologien für die Genuss-Märkte.* Zukunftsinstitut-Studie. Kelkheim.

Kisabaka, L. (2001). *Marketing für Luxusprodukte* (Vol. 32). Dissertation, Fördergesellschaft Produkt-Marketing, Cologne.

Kröber-Riel, W. und Gröppel-Klein, A. (2013). Konsumentenverhalten. 10. Auflage, Verlag Franz Vahlen GmbH, München.

Lasslop, I. (2005). Identitätsorientierte Führung von Luxusmarken. In H. Meffert, C. Burmann, M. Körs (Hrsg.), *Markenmanagement* (S.469-494). 2. Auflage, Betriebswirtschaftlicher Verlag Dr. Th. Gabler/ GWV Fachverlage GmbH, Wiesbaden.

Leitzmann, C. (1993). Food Quality-Definition and a Holistic View. In H. Sommer, B. Petersen, & P. v. Wittke (Eds.), *Safeguarding Food Quality* (pp. 3-15). Springer Verlag Berlin Heidelberg,

Liebel, F. (2007). Motivforschung: Eine kognitionspsychologische Perspektive. In G. Naderer, & E. Balzer (Hrsg.), *Qualitative Marktforschung in Theorie und Praxis* (S. 451-468). Betriebswirtschaftlicher Verlag Dr. Th. Gabler/ GWV Fachverlage GmbH, Wiesbaden.

Lüth, M. (2005). *Zielgruppensegmente und Positionierungsstrategien für das Marketing von Premium-Lebensmitteln.* Dissertation, Georg- August- Universität Göttingen.

Max Rubner-Institut (2008). *Nationale Verzehrsstudie II.* Ergebnisbericht Teil 1, Die bundesweite Befragung zur Ernährung von Jugendlichen und Erwachsenen, Karlsruhe.

Meurer, J. (2012). Ebony or Ivory–wie glänzend ist die Zukunft des Luxus in Deutschland? Kritische Reflexionen zum Luxusmarkenmanagement. In C. Burmann, V. König, & J. Meurer (Hrsg.), *Identitätsbasierte Luxusmarkenführung* (S. 321-336). Springer Fachmedien Wiesbaden.

Meurer, J., & Manninger, K. (2012). Quo vadis globale Luxusmarkenführung–Status, Trends und Top-Themen für die CMO-Agenda. In C. Burmann, V. König, & J. Meurer (Hrsg.), *Identitätsbasierte Luxusmarkenführung* (S. 321-336). Springer Fachmedien Wiesbaden.

Nestlé Deutschland AG (2012) (Hrsg.). *Nestlé Studie 2012, Das is(s)t Qualität*, Auszüge aus der Nestlé Studie 2012.

Nielsen (2014). Umsatz mit Feinkost im Lebensmitteleinzelhandel in Deutschland nach Warengruppen im ersten Halbjahr 2014 (in Millionen Euro). URL: http://de.statista.com/statistik/daten/studie/218578/umfrage/umsatzentwicklung-der-feinkost-warengruppen-in-deutschland/. Zugriff am 20. Oktober 2014.

Nitzko, S., & Spiller, A. (2014). Zielgruppenansätze in der Lebensmittelvermarktung. In M. Halfmann (Hrsg.), *Zielgruppen im Konsumentenmarketing* (S. 315-332). Springer Fachmedien Wiesbaden.

Padilla Bravo, C. A. (2013): Consumer Behaviour towards Organic Food and Performance of Certification Standards. Dissertation, Georg-August-Universität Göttingen, Cuvillier Verlag, Göttingen.

Padilla Bravo, C., Cordts, A., Schulze, B., & Spiller, A. (2013). Assessing determinants of organic food consumption using data from the German National Nutrition Survey II. *Food Quality and Preference, 28*(1), 60-70.

Panayides, P. (2013). Coefficient Alpha: Interpret With Caution. *Europe's Journal of Psychology, 9*(4), 687-696.

Prüne, G. (2013). *Luxus und Nachhaltigkeit.* Springer Fachmedien Wiesbaden.

SGS (2014). Vertrauen und Skepsis: Was leitet die Deutschen beim Lebensmitteleinkauf? SGS-Verbraucherstudie 2014: Ergebnisse einer bevölkerungsrepräsentativen Befragung. Hamburg: SGS Germany GmbH.

Silverstein, M. J., Fiske, N., & Butman, J. (2008). *Trading Up: why consumers want new luxury goods--and how companies create them.* Penguin.

Spiller, A. (2010). *Marketing Basics, Ein Online-Lehrbuch.* Georg-August-Universität Göttingen.

Statistisches Bundesamt (2013a) (Hrsg.). Bevölkerung 1976 bis 2012 nach Bildungsabschluss. In Bildungsstand der Bevölkerung, Wiesbaden.

Statistisches Bundesamt (2013b). Zahlen & Fakten URL: https://www.destatis.de/DE/ZahlenFakten/GesellschaftStaat/StaatGesellschaft.html;jsessionid=C96AEA9A9BCDEF2DAA17DF50700EB5A6.cae1. Zugriff am 31. Januar 2013.

Van der Veen, M. (2003). When is food a luxury?. *World Archaeology, 34*(3), 405-427.

Veblen, T. (1899). *The theory of the leisure class*. Oxford University Press.

Wiedmann, K. P., Hennigs, N., & Siebels, A. (2007). Measuring consumers' luxury value perception: a cross-cultural framework. *Academy of Marketing Science Review, 7*(7), 333-361.

Wiedmann, K. P., Hennigs, N., & Siebels, A. (2009). Value-based segmentation of luxury consumption behavior. *Psychology & Marketing, 26*(7), 625-651.

Yeoman, I. (2011). The changing behaviours of luxury consumption. *Journal of Revenue & Pricing Management, 10*(1), 47-50.

Yeoman, I., & McMahon-Beattie, U. (2006). Luxury markets and premium pricing. *Journal of Revenue and Pricing Management, 4*(4), 319-328.

Zeithaml, V. A. (1988). Consumer Perceptions of Price, Quality, and Value: A Means-End Model and Synthesis of Evidence. *Journal of Marketing*, Vol. 52 (July 1988), pp. 2-22.

Zühlsdorf, A., & Spiller, A. (2012). *Trends in der Lebensmittelvermarktung. Begleitforschung zum Internetportal lebensmittelklarheit.de: Marketing theoretische Einordnung praktischer Erscheinungsformen und verbraucherpolitische Bewertung*. Hrsg.: Agrifood Consulting GmbH Spiller, Zühlsdorf + Voss, Göttingen.

II.2 Segmentation of German Consumers Based on Perceived Dimensions of Luxury Food

Authors: **Laura Hartmann, Sina Nitzko, Achim Spiller**

Georg-August-University of Goettingen

For submission.

Abstract

Recent trends in the German food sector have shifted consumption motives segmentally toward a focus on indulgence, quality and sustainability. At the same time, there is evidence of a change in Western motives for luxury consumption, shifting emphasis to hedonism, self-actualization and quality instead of prestige and conspicuous consumption. Those parallel developments give rise to the question of how is food related to the concept of luxury. Based on general marketing research findings, first this study investigates the perceived dimensions of luxury food. Second, consumer segments are built upon. By means of a principal component analysis and a three-stage cluster analysis, seven dimensions are revealed, which generate four consumer segments. The results show that six of the perceived dimensions of luxury food correspond to those that are found for general luxury. Meanwhile, sustainability and authenticity form a further dimension, mirroring new luxury values. In the cluster analysis, two segments can be identified as target groups for classical premium foods, while one is predominated by new luxury values and one is expected to buy lower quality food.

Keywords

Luxury, Food, Consumer Behavior, Segmentation

Introduction

Considerations on general luxury consumption in Western cultures

Recently, international luxury markets have been flourishing, and researchers predict a further upward trend (Bellaiche et al., 2012). In 2012, the global luxury market volume was estimated at 1.42 trillion Euros: 14.9% of this was based on personal luxury goods like apparel, accessories and perfumes and 46.8% on so-called *luxury experience* like activities of auction houses, hotels and exclusive trips. A 38.3% share belongs to other luxury goods, where the category of food, wines and spirits, for example, contributes to 5.4% (76 billion Euros in market volume) (Müller-Stewens, 2013). The average annual growth rate of the market for personal luxury goods, luxury cars and experiential luxury between 2010 and 2012 was approximately 13% (Bellaiche et al., 2012). This trend of increasing activities on luxury markets is also confirmed for Germany, which is one of the most significant countries of origins for luxury brands in the international context (Meurer, 2012).

In general, the literature considers luxury as a multi-faceted construct (Wiedmann, Hennigs, & Siebels, 2007). The original Latin term *luxus* stands for extravagance, indulgence, affluence and amplitude (Glare, 1992). Furthermore, the "dream value", which signifies the strong desire to consume a specific good, was revealed as a decisive characteristic of luxury (Albrecht et al., 2013; Dubois & Paternault, 1995). Even though, the definition of those goods means luxury for an individual person is highly subjective. For example, it can be differentiated between material and immaterial luxury values (Pflanz, 2004). In general, the description of the construct luxury is often based on consumption motives. They are not only influenced by culture, but they are also closely linked with a change in value systems over time (Wong & Ahuvia, 1998).

Motives for luxury consumption can generally be subdivided into three categories (Table 1). These are externally-related, internally-related and finally, motives like price and quality, which are hybrids in this context. Motives for the first category gain importance only in the environment of social structures. They assume the comparison with fellow men and primarily describe the intention to signal social status and milieu affiliation (Han, Nunes, & Dreze, 2010; Wiedmann, Hennigs, & Siebels, 2009). This is closely linked to the concept of *Conspicuous Consumption* described by Veblen (1899). Prestige, status, tradition and self-presentation are further externally-oriented motives that are discussed in the literature (Ascheberg, Meurer, & Oesterling, 2012; Vigneron & Johnson, 2004). Mason (1993) correspondingly describes three different social effects induced by luxury consumption: the

126

Veblen effect, the snob effect and the bandwagon effect. The Veblen effect refers to the demonstration of wealth, while the Snob effect covers the demonstration of exclusiveness and social grandeur. The Bandwagon effect emphasizes the membership to a particular social class or group. The Snob effect increases when differences in income within a society decrease, while the Veblen effect is associated with an upward mobility within a society. The latter is an expression of success, material possession and social advancement, and describes the positive correlation between demand and price of a luxury good (Veblen, 1899).

Internally-related motives for luxury consumption result from the desire for indulgence and self-realization. Here, the consumption of luxury goods is the means by which people pursue hedonic aims, and where they live out and refine their personality and individuality (Meurer, 2012). The value of a luxury good is generated from an internal personal sensation or emotional gain. It is therefore uncoupled from the embeddedness of consumers within their social environment (Atwal & Williams, 2009). The *luxury* experience as well as the satisfaction of subjective claims and desires, like striving for perfection, are examples for addressing these patterns of consumption in literature (Atwal & Williams, 2009; Meurer, 2012). Yeoman and McMahon-Beattie (2006) describe a trend toward luxury consumption as an instrument for living a more pleasurable life, the fulfillment of the desire for authenticity, experience and abandonment. Tsai (2005) empirically finds that the personal value of luxury consumption is positively related to the repurchase intention for a luxury brand. Kisabaka (2001) discusses the marked tendency of luxury consumers toward hedonism and identifies, among other things, the need for self-realization as a decisive motive for luxury consumption.

The third category of hybrid drivers for luxury consumption is based on the product characteristics of quality, usability, uniqueness, materialism and price. They have different functions with regard to the fulfillments of externally-related or internally-related consumption motives, depending on the individual consumer. A general categorization of these attributes into the former two categories is hence not possible. People associate usability, quality, uniqueness and materialism as well as specific price policies with the value of luxury products (Fassnacht, Kluge, & Mohr, 2013; Hornig, Fischer, & Schollmeyer, 2013; Wiedmann, Hennigs, & Siebels, 2009). At the same time, they can serve as instruments to not only signal status and prestige but to additionally satisfy personality-based (hedonistic) desires (Vigneron & Johnson, 1999; Yeoman & McMahon-Beattie, 2006).

Table 1. Three categories of motives for luxury consumption

Externally-related motives	Internally-related motives	Hybrid motives
Conspicuous consumption/Signaling	Hedonism and indulgence	Quality
Social stratification	Experience	Usability
Status and prestige	Self-realization	Uniqueness
Tradition	Self-fulfillment	Materialism
Self-portrayal	Individuality	Price
	Inspiration	
	Authenticity	

In recent times, a change in the motives of Western luxury consumption has been discussed. Traditional consumption patterns like prestige, social stratification and conspicuousness, as described by Veblen (1899), take a backseat to modern motives like hedonism, immaterialism and sustainability. Consuming expensive brands and services or accumulating items of property does not primarily serve the externally-related self-portrayal. Instead, consumers increasingly satisfy their internal longing by means of consuming luxury. The demand for hedonistic actions and individuality becomes a superficial motive for a delightful, luxurious lifestyle in an emancipated affluent society (Ascheberg, Meurer, & Oesterling, 2012; Atwal & Williams, 2008; Vigneron & Johnson, 2004; Yeoman & McMahon-Beattie, 2012; Yeoman, 2011).

As material luxury is widely accessible across social layers in industrial nations, intrinsic desires in contrast to social desires could possibly gain importance (Yeoman & McMahon-Beattie, 2006). In this context, Kisabaka (2001) speaks of a value shift that put emphasis on self-directed and stimulating values. Due to this, they turn into decisive factors on the market for luxury goods. They provide a cognitive framework for the desire for luxury and displace economic and social purchasing criteria as parameters of explanation for luxury customers' behavior. Luxury goods are assessed less by their physical value but more by the basis of the added emotional value that consumers associate with them (Mostovicz, 2010). Mohr (2014) analyzes the modern paradigm of diagonal social integration. He describes its aim to become an example for the personal social environment by means of living a happy and successful life, and is therefore directly related to the shift in consumption motives in Western societies. The differing horizontal social integration, where status symbols and demonstration of wealth are used to signal the membership to higher social classes, becomes a less attractive societal aim. However, observing the developments in motives for luxury consumption reveals that the traditional assumption of purely prestige-minded and materialistic target groups in the market for luxury goods cannot be supported any longer.

Developments in food consumption in Germany

In accordance with the findings for luxury consumption, there is an observable shift in food consumption motives for some German consumer segments. Product characteristics like quality, sustainability and authenticity, as well as hedonistic values like food enjoyment have gained importance, while low prices are decreasingly relevant for purchase decisions (Nestlé Deutschland AG, 2012; Nitzko & Spiller, 2014; SGS, 2014; Page, 2006). At the same time, studies reveal that the general significance of nutrition, common meals with family and friends, cooking and enjoying eating in good restaurants is segmentally high (Brunso, Fjord, & Grunert, 2002). Brunso, Grunert, and Bredahl (1996) identify "the Adventurous Food Consumers" (AFC) who represent one of five cross-cultural segments regarding the Food-Related Lifestyle. The segments build upon five dimensions: ways of shopping, cooking methods, quality aspects, consumption situations and purchasing motives. The AFC likes to cook, is concerned about quality and associates eating with self-fulfillment. This consumer type, 24% of the German population in 1996, places a high priority on food quality and is generally likely to buy gourmet food (Brunso, Grunert, & Bredahl, 1996; Wycherly, McCarthy, & Cowan, 2008). Lueth, Spiller, and Enneking (2004) further identify two nutrition-conscious segments within Germany: first, people who like to cook and are mainly concerned about health and regionalism of food, and second, those who like to eat in good restaurants, who are health- and fitness- oriented, and prefer organic, regional or Fair Trade food. Correspondingly, Kirig and Ruetzler (2007) describe recent food trends in Germany as shifting consumption motives from the former mentality of avariciousness toward health consciousness, quality, hedonistic experience and luxury, and, in accordance to the paragraph above, they postulate that general pleasure markets will be influenced by new luxury values like moral enjoyment, authenticity and health. The authors define food styles as mirrors for trends in society and as indicators for motivational changes in other branches. An explanation for this may be given by the general embedment of food styles into individual value systems. Brunso, Scholderer, and Grunert (2004) found that ten motivational domains proposed by

Schwartz (1992), covering inter alia hedonism, stimulation and self-direction, are significantly related to the way how people consume food.

Summarily, the substantially growing demand recognized for organic, regional and Fair Trade food products can be taken as a sign for the establishment of less price-sensitive but more quality-, indulgence-, health- and sustainability-oriented food styles (Baker et al., 2004; Honkanen, Verplanken, & Olsen, 2006; Institut für Demoskopie Allensbach, 2014; Magnusson et al., 2003).

Those identified developments shed new light on luxurious food products. Major intersections between motives for luxury consumption and food consumption imply that food can be associated with luxury and vice versa. Indulgence and quality seem to be in the spotlight of significant consumer segments, presenting a challenge for food marketing. The question appears if marketing policies in the food sector now should rather transport intrinsic values than price leadership as has been the aim for a long time.

Based on the shift in consumption patterns for luxury in general as well as the upcoming trend for qualitative, conscious food styles, this study investigates the perceived dimensions of luxury food. Building upon the revealed dimensions, consumer segments are identified in order to define target groups for the luxury food market. The overall aim is to understand what actually are the crucial characteristics of luxury food and how consumers can be segmented based on what they associate with this expression.

Literature review and construct definition

Considerations of luxury food

Approaches to answer the question what luxury food means can be found from an archaeological, historical point of view in theoretical literature. Van der Veen (2003) characterizes luxury food first, as offering a refinement of a food that is widely desired and second, by means of distinction. She bases the latter on either quantity or quality, symbolizing either success and prestige or exclusivity and distance. The author further concludes that there are no specific items of food universally considered to be a luxury; it depends rather on a particular place, time and society. In developing countries for example, food quantity is used as a luxury in order to gain prestige, while in complex societies, the emphasis is on food quality for creating exclusivity. This approach is inspired by Berry (1994) who defines luxury in general as objects of desire, as contradiction to the norm and instrumental needs. Luxury

130

means a refinement or specific quality of a need, e.g. fresh bread instead of bread for satisfying hunger. Needs and desires are not stable over time, they only can be defined in relation to each other and therefore depend on time, stage of development in a society and social layer. Hayden (2003) also addresses the dependence on time stating, "In short, our eating habits today largely are the result of, and reflect, the luxury foods of the past." (p. 459). Van der Veen (2003) further reveals a social component of luxury food as it serves as an instrument for establishing and tightening social relations. Therefore, it is referred to as the meaning of food in feasts. Following Dietler (1990, 1996), four categories of feasts are described where food plays different roles in order to either symbolize exclusiveness or distance, social power or to build up social bonds.

Other contributions toward luxury food are wine studies. Other contributions toward luxury food are studies on wine. Beverland (2004) states that wine is a suitable example where research on luxury can be done because of three reasons. First, auction prices have increased significantly; second, the evaluation of quality and style are individual and highly variable; and third, premium wine brands, which are defined by means of the product characteristic price, are not marketed by mass marketing. There, three indirect definitions of luxury are found and applied in the context of food: price, quality and marketing. These characteristics defining luxury food may remind of how classical delicacies and premium food are marketed. Following on these results, Beverland (2005) uses wine as a basis and investigates another luxury dimension which is authenticity. This is here highlighted as a core component of brand management that becomes increasingly important. Luxury wines are associated with authenticity because they represent the oldest existing brands and are marketed by techniques distanced from industrial products and commerce. Luxury wines are defined in this paper by price and quality. The author investigates the definition of authenticity and concludes that it is "projected via a sincere story that involves the avowal of commitments to traditions, passion for craft and production excellence, and the public disavowal of the role of modern industrial attributes and commercial motivation" (p. 1025). Quality, origin and the decoupling from mass production are found as strong definitional attributes of authenticity. Accordingly, Beverland (2006) highlights heritage and pedigree, stylistic consistency, quality commitments, relationship to place, method of production, and downplaying commercial motives as the six attributes of authenticity, by means of research on wineries and wine consumers. Sidali and Hemmerling (2014) show that authentic food products are associated with more traditional and nostalgic production methods, in contrast to production methods that result from industrialized and intensified agriculture. Furthermore, they find that consumers perceive authenticity as an attribute of small-scale, artisan-produced foods. The image of the producer as an underprivileged actor in niche food markets, in contrast to global food brands, is thereby important for the perceived definition of authentic food. Summarily, origin and the rejection of conventional mass production methods are found again to define authenticity.

Kemp (1998) performs three empirical studies answering the question of whether some food items are perceived as luxury items. Respondents were asked to rate, among other items, wine, bars of chocolate, lunch with friends, fruit juice, fish, milk, bread, a favorite pie in the fridge and champagne on a necessity-luxury scale (with 1=complete necessity and 9=complete luxury). The differentiation of needs and desires done by Berry (1994) inspires the scale. The results reveal that champagne is associated with luxury most often (rated 8 on average), followed by wine, a bar of chocolate and lunch with friends (rated 7 on average), while fruit juice, fish and a favorite pie in the fridge are third-ranked (rated 5 on average), milk is fourth-ranked (rated 3 on average) and bread is least associated with luxury (rated 2 on average). On the one hand, these results confirm the assumption that wine has a high perceived luxury value. On the other hand, the item "favorite pie in the fridge" calls into consideration a new aspect of luxury, which is based on the consumer's personality, namely on self-directed consumption motives. In conclusion, different aspects of perceived luxury food are found in these studies. The latter named mirrors individual luxury dimensions while champagne, wine and chocolate can be assumed as representatives of quality, price and uniqueness.

Construct definition

The literature mentioned above shows that there is a change in patterns of Western luxury consumption, bringing out an increased importance of personal-related and self-directed motives. Furthermore, the literature reveals a parallel change of motives in food styles resulting in a focus on quality, sustainability and indulgence. With reference to these developments, which imply a definition of luxury food based on a catalogue of modern and classical consumption motives, the concept of this study is chosen to be based on a motivational, multidimensional approach. Wiedmann, Hennigs, and Siebels (2007, 2009) inspire the structural model for this study by proposing a framework for the definition of a general luxury value. It is extended here by the dimension of sustainability and authenticity.

This results in five investigated dimensions of luxury values where authenticity and sustainability are chosen to present new luxury attributes in the context of food (Figure 1). In accordance with Jarvis, MacKenzie, and Podsakoff (2003), the mode of measurement is defined to be formative. The causality is from the indicators to the first- and second-order latent variables, which means that the attributes of a food product determines if customers associate a luxury value with it.

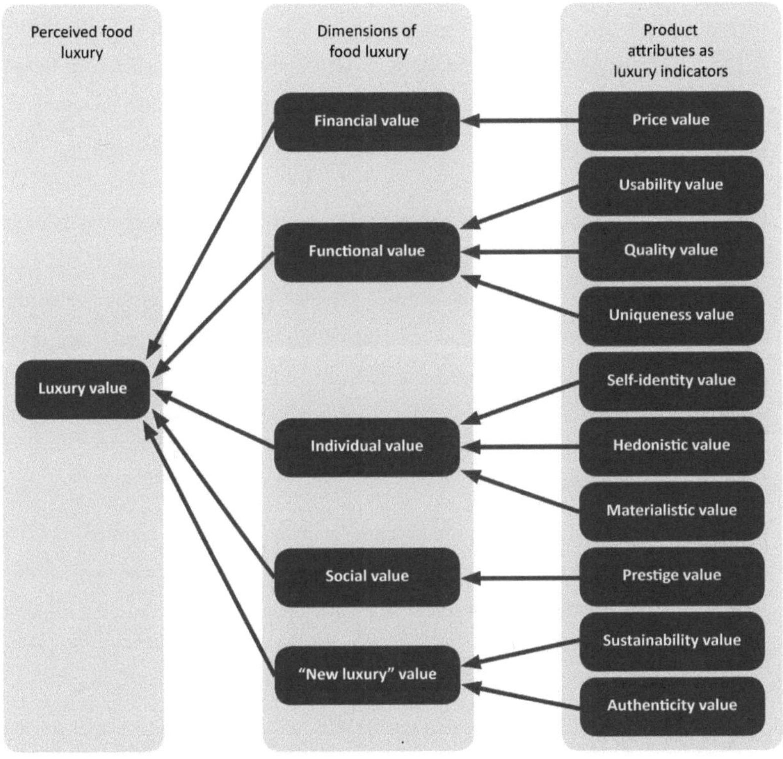

Figure 1. Basic model for investigations[17]

---

[17] Inspired by Wiedmann, Hennigs, and Siebels (2007, 2009).

The financial, functional, individual and social values are associated with general luxury, as the literature repeatedly shows. High prices are closely connected with perceived quality and exclusivity, and they ultimately act as a moderator variable for the determination of perceived luxury (Erickson & Johansson, 1985; Fassnacht, Kluge, & Mohr, 2013; Hornig, Fischer, & Schollmeyer, 2013; Serraf, 1991; Wheatley & Chiu, 1977; Wiedmann, Hennigs, & Siebels, 2007, 2009). Usability, quality and uniqueness which present the dimension of functionality are revealed to be characteristics of luxury products (Kisabaka, 2001; Vigneron & Johnson, 2004; Wiedmann, Hennigs, & Siebels, 2007, 2009; Yeoman & McMahon-Beattie, 2006) as well as the individual dimension in terms of self-identity, hedonism and materialism (Dubois & Laurent, 1994; Hudders & Pandelaere, 2012; Puntoni, 2001; Vigneron & Johnson, 2004; Wiedmann, Hennigs, & Siebels, 2007; 2009), and the social value in terms of prestige (Mason, 1993; Vigneron & Johnson, 1999, 2004; Wiedmann, Hennigs, & Siebels, 2007, 2009).

With regard to the question of what attributes do consumers perceive as adding value to luxury foods, the dimensions defined for general luxury seem to be applicable. The literature on luxury mentioned above gives evidence for the existence of all four value dimensions: price, quality, self-identity and social meaning. At the same time, the described new emphasis on quality and indulgence in significant German consumer segments reveals the perceived importance of functional and individual values of food. According to these findings, it is hypothesized:

H1: The conceptual framework proposed by Wiedmann, Hennigs, and Siebels (2007, 2009), which includes a financial, functional, individual and social dimension of luxury, can also serve as a basis for defining consumer segments upon the perceived dimensions of luxury food.

Sustainability and authenticity are two crucial attributes of food products that consumers have recently demanded (Baker et al., 2004; Honkanen, Verplanken, & Olsen, 2006; Institut für Demoskopie Allensbach, 2014; Magnusson et al., 2003; Padilla Bravo et al., 2013). Concurrently, both dimensions are actively discussed in conjunction with luxury at all. Based on a general rise of sustainability concerns in consumer behavior, it is seen as the challenge for luxury marketing to address concepts like environment friendliness, animal welfare and social sustainability (Bilharz & Belz, 2008; Blevis et al., 2007; Joy et al., 2012; Kapferer, 2010).

Authenticity is known to be a crucial instrument in order to create a unique brand identity, especially in the context of luxury, where authenticity is a core element of brands, symbolizing uniqueness, exclusiveness and quality (Beverland, 2005 & 2006; Schallehn, 2012). Meurer and Manninger (2012) even postulate it as one of the "Mega-Trends" within the international luxury branch. Consequently, the second hypothesis of this paper is:

H2: Sustainability and authenticity form an additional dimension that is significant for the definition of luxury food.

Methodology

The definition of luxury value in the context of food is investigated by means of the above-introduced structural model, which Wiedmann, Hennigs, and Siebels (2007, 2009) inspired. Accordingly, an explorative factor analysis (principal component analysis with varimax rotation method) is applied first to identify the relevant luxury dimensions. Second, consumer segments are identified with a three-stage cluster analysis, which is based on the revealed factors. The database was generated in the summer of 2014 via an online questionnaire received from 36 German respondents.[18] The first three items with the highest loadings of each of the factors revealed by Wiedmann, Hennigs, and Siebels (2009)[19, 20] were adapted to the food context and translated into German. For the value of self-directed pleasure, the item "I never buy a luxury food product inconsistent with my own taste." was added. Three further items each were added in order to address price, sustainability and authenticity (Table 3). Respondents evaluated all items on a five-point Likert scale ranging from full agreement (+2) to no agreement at all (-2). A second part investigates the luxury food purchasing behavior. Therefore, respondents provided information on their consumption frequency for 21 food items. In this context, a five-point scale from frequent consumption to no consumption at all was applied. The items were chosen to represent different luxury dimensions stated by Wiedmann, Hennigs, and Siebels (2007, 2009) (Table 4). A second explorative factor analysis

---

[18] Nine hundred thirty-six respondents remained after data cleansing. The original sample was 1050 respondents.

[19] With limiting the online questionnaire to the first three items with the highest loadings each, we adapted to time-constraints.

[20] Wiedmann, Hennigs, and Siebels (2009) adopted existing and tested measures (e.g. Dubois & Laurent, 1994; Richins & Dawson,1992; Tsai, 2005) and added items that correspond to results from explanatory interviews.

(principal component analysis with varimax rotation method) reveals their latent structure. These factors as well as additional information on the perceived importance of luxury food, willingness to pay, food styles and preferred food shopping venues as well as socio-demographic data serve as clusters describing variables by cross-tabulation. The sample was chosen to be representative for the German population with respect to gender, age, state, net household income and the highest school leaving qualification of the main earner of the household. A pretest of 25 respondents, including research experts and others, was conducted in advance. For the statistical estimations, IBM SPSS Statistics 22 was the software used.

Sample description

Table 2 shows the socio-demographic data that were used as a base for representativeness in comparison to respective distributions within Germany. It is gender, age, region, net household income and the highest school leaving qualifications of the main earner among household members.

Table 2. Comparison between socio-demographic data in the sample and the German population

Socio-demographic data	Sample[1]	German population[2]
Gender	Female: 51.8%% Male: 48.2%	Female: 51.0% Male: 49.0%
Age	18–29 years: 16.5% 30–39 years: 15.2% 40–49 years: 20.9% 50–59 years: 22.0% 60–69 years: 18.3% 70–75 years: 7%	18–29 years: 18.7% 30–39 years: 15.9%% 40–49 years: 21.3%% 50–59 years: 20% 60–69 years: 14.7% 70–75 years: 9.4%

Region[3]	North: 18.5% East: 17.5% West: 35.5% South: 28.5%	North: 18.2% East: 17.7% West: 35.5% South: 28.6%
Net household income per month	Less than 900€: 8.2% 901–1300€: 11.2% 1301–1500€: 5.7% 1501–2000€: 14% 2001–2600€: 13.8% 2601–3600€: 12.8% 3601–5000€: 10.9% 5001–7600€: 6.3% 7601–9000€: 3.1% 9001–12600€: 1.5% 12601–18000€: 0.9% More than 18000€: 0.7% No comment: 10.9%	Less than 900€: 8.7% 901–1300€: 11.5% 1301–1500€: 5.8% 1501–2000€: 14.7% 2001–2600€: 14.4% 2601–3600€: 17.3% 3601–5000€: 14.6% 5001–18000 Euro: 13.1%

| **Highest school leaving qualification of the main earner among household members** | Without school leaving qualification: 0.7%

Primary school/lower secondary school: 22.1%

Qualified degree of lower secondary school: 13.8%

Secondary school: 25.4%

Polytechnic high school: 6.1%

Technical school: 4.0%

General or specialized higher education (Abitur): 25.3%

Others: 1.7%

No comment: 0.9% | Lower Education (primary school, lower secondary school, qualified degree of lower secondary school): 35.6%

Middle Education (secondary school, polytechnic high school): 29.1%

Higher Education (technical school, general or specialized higher education): 27.3% |

[1]N=936.

[2]rounded values; source: Statistisches Bundesamt (2013).

[3]North: Schleswig-Holstein, Hamburg, Mecklenburg-Vorpommern, Lower Saxony, Bremen; East: Berlin, Brandenburg, Saxony-Anhalt, Saxony, Thuringia; West: North Rhine-Westphalia, Hesse, Rhineland-Palatinate, Saarland; South: Bavaria, Baden-Wuerttemberg.

For revealing the general attitude toward luxury food, participants were asked first, how important luxury is for them in the context of food. Nineteen and eight-tenths percent answered that they found it very important or important, 35.2% marked the intermediate category and 45% find it unimportant or very unimportant. Second, the respondents were asked to evaluate the statement "Generally, I'm turned towards luxury with foods.". Regarding this, 22.2% agreed or totally agreed, 31% were indecisive and 46.9% disagreed or totally disagreed. Finally, the question was asked if participants were willing to pay higher prices for food items that they perceived to have some luxury value. The results show that 35.1% were willing, 37.1% were not sure, and 27.8% were not willing.

These results could imply that social desirability plays a role when addressing the topic of luxury food (Grimm, 2010). High shares of participants who chose the intermediate category of answers could reflect uncertainty in how to evaluate the luxury topic. Furthermore, high shares of people who chose negative answers to the first two questions are not consistent with the result that more people are willing to pay a higher price for luxury food than people who are not willing to pay a higher price. Since luxury generally has polarizing power, inducing a strong aversion in some consumer segments, participants could be kept away from truly revealing their attitude (Allérès, 1993; Dubois & Laurent, 1994).

Explorative factor analysis of luxury food dimensions

The first factor analysis of luxury motives revealed seven factors that combine the different luxury dimensions from the basic model (Figure 1). Seven items were deleted because their loadings were lower than 0.5. This rule is stricter than suggested in the literature (Fabrigar, Wegener, MacCallum, & Strahan, 1999; Ford, MacCallum, & Tait, 1986). The Kayser-Meyer-Olkin criterium was 0.889, which is higher than a proposed minimum value of 0.6 (Robinson, Shaver, & Wrightsman, 1991). The explained overall variance amounts to 66.086%. Three factors show a Cronbach's Alpha higher than 0.85 and three one higher than 0.65. One is 0.262. All values are located in the desirable range or close to it (Bland & Altman, 1997; Nunnally, 1978). The factors are structured as shown in Table 3.

Table 3. Explorative factor analysis of attitudinal statements[1]

Factor 1: Prestige and Hedonism (Cronbach's Alpha: 0.902, explained share of overall variance: 16.006%)	Factor loading	μ ; σ
I like to know what food products and food brands make good impressions on others.	0.818	-1.11; 0.997
I usually keep up with food style changes by watching what others buy.	0.780	-0.93; 1.061
Before purchasing a food product it is important to know what food brands or products make good impressions on others.	0.777	-1.10; 1.109
Purchasing luxury food provides deeper meaning in my life.	0.726	-1.02; 1.049
Self-actualization is an important motivator for my luxury food consumption.	0.724	-0.79; 1.117
When I consume luxury food, cultural development is an important motivator.	0.718	-0.76; 1.022
When in a bad mood, I may consume luxury food for alleviating the emotional burden.	0.606	-0.72; 1.123
Factor 2: Self-identity and Quality (Cronbach's Alpha: 0.791, explained share of overall variance: 11.000%)	**Factor loading**	**μ ; σ**
I'm inclined to evaluate the substantive attributes and performance of luxury food products myself rather than listen to others' opinions.	0.772	1.18; 0.903
A luxury food product preferred by many people but that does not meet my quality standards will never enter into my purchase consideration.	0.738	1.19; 0.957
Luxury food I buy must match my lifestyle and me.	0.710	0.91; 1.045
I never buy a luxury food product inconsistent with my own taste.	0.702	1.27; 0.982
I buy luxury food for satisfying my personal needs without any attempt to make an impression on other people.	0.668	0.85; 1.227
It is my own desire when I buy a luxury food product, not meeting the expectations of others.	0.565	1.16; 0.946
Factor 3: Sustainability and Authenticity (Cronbach's Alpha: 0.857, explained share of overall variance: 9.763%)	**Factor loading**	**μ ; σ**
Luxury food products should be produced environmentally friendly.	0.820	0.68; 1.096
I only buy a luxury food product if it is produced under fair conditions for humans and animals.	0.813	0.27; 1.163
It is important to me that luxury food products I buy are organic.	0.802	0.03; 1.178
I like to know where the luxury food products I buy are produced.	0.781	0.56; 1.137

Factor 4: Materialism (Cronbach's Alpha: 0.891, explained share of overall variance: 9.045%)	Factor loading	μ ; σ
It sometimes bothers me quite a bit that I can't afford to buy all the luxury food products I'd like.	0.859	-0.58; 1.270
I'd be happier if I could afford to buy more luxury food products.	0.842	-0.68; 1.188
My life would be better if I bought certain luxury food products I don't consume.	0.789	-0.86; 1.111
Factor 5: Usability (Cronbach's Alpha: 0.759; explained share of overall variance: 7.879%)	Factor loading	μ ; σ
In my opinion, luxury foods are really useless.[2]	0.849	-0.02; 1.117
In my opinion, luxury foods are swanky.[2]	0.824	0.34; 1.133
In my opinion, luxury foods are pleasant.	0.591	-0.10; 1.115
Factor 6: Uniqueness (Cronbach's Alpha: 0.626; explained share of overall variance: 6.322%)	Factor loading	μ ; σ
True luxury food cannot be mass-produced.	0.793	0.81; 1.001
A luxury food product cannot be sold in supermarkets.	0.751	-0.03; 1.116
Few people consume true luxury food.	0.678	0.74; 1.020
Factor 7: Price (Cronbach's Alpha: 0.651; explained share of overall variance: 6.071%)	Factor loading	μ ; σ
I rather buy a certain food product when its price is comparatively high.	0.805	-0.52; 1.034
When buying food, the price is not decisively.[2]	-0.747	0.38; 1.050
I associate high prices for food products with particularly high quality.	0.573	-0.08; 1.087

[1] Items evaluated with five-point Likert scale (from -2=full disagreement to 2=full agreement); principal component analysis with varimax rotation and Kaiser normalization; 7 iterations, Kayser-Meyer-Olkin criterion: 0.889; Bartlett significance: 0.000; cumulated explained share of overall significance: 66.086%.

[2] Reverse coded items.

The first factor, *Prestige and Hedonism*, includes items from the social and individual dimension in the initially stated conceptual model. The prestige component expresses a desire to impress others and be informed about food trends. It is thus found to be an analogy to the significant social meaning of food generally and luxury food in particular (Appadurai, 1988; Gosden & Hather, 2004; Gumermann, 1997; Hayden, 2003; Van der Veen, 2003). Food consumption is often a social ritual, e.g. in feasts, where the choice and preparation of particular food items as well as the way to serve and eat them symbolize identification and intention within a social environment (Lupton, 1994). The hedonic component in this factor refers to *Self-gift Giving, Life Enrichment* and *Extravagance* that Wiedmann, Hennigs, and Siebels (2009) propose. The first factor relates indulgence and self-realization by means of luxury food consumption with sociability in terms of prestige-based motives. The enjoyment of luxury food (hedonism) and the desire to impress others or to meet current food trends (prestige) seem to facilitate each other or even to be mutually dependent. Well-being and indulgence, and dealing with a social environment are probably more related to each other according to food consumption than in general luxury consumption. That being said, Vigneron and Johnson (1999) also define a perceived hedonic value as one of four crucial characteristics that distance general prestige brands from others.

Secondly, *Self-identity and Quality* are summarized in a factor consisting of seven items. In accordance with the findings for general luxury (Kapferer & Bastien, 2008; Vigneron & Johnson, 2004; Wiedmann, Hennigs, & Siebels, 2007, 2009, Yeoman & McMahon-Beattie, 2009), both motives obviously play a significant role in the definition of luxury food. The mean values of evaluation are the highest among all factors. Interestingly, people seem to relate both motives with regard to luxury food while quality and self-identity formed separated factors with regards to general luxury (Wiedmann, Hennigs, & Siebels, 2009). Similar to the first factor, it can be shown that two typical luxury values are perceived to be mutually dependent, suggesting that the level of quality and the content of self-identity positively influence each other.

The third factor *Sustainability and Authenticity* represents a hypothesized new luxury dimension. The highest loading is estimated for the organic food aspect. Authenticity refers first to the last item "I like to know where the luxury food products I buy are produced", mirroring the aspect of *origin*. Second, it is indirectly referred to by means of the items for sustainability: According to Beverland (2006) as well as Sidali and Hemmerling (2014), the disavowal of industrialized and mass production methods is one important aspect in the definition of authenticity.

The formation of this factor mirrors, on the one hand, the general trend in industrialized countries toward sustainable food consumption with respect to whole food, regionality, fair trade and animal welfare (Harper & Makatouni, 2002; Institut für Demoskopie Allensbach, 2014). On the other hand, it confirms the general assumption that luxury in Western cultures is increasingly associated with sustainability and authenticity (Beverland, 2005, 2006; Bilharz & Belz, 2008; Joy et al., 2012; Kapferer, 2010).

The fourth factor contains all items having due with the *materialistic value of luxury.* Similarly to the first and second factor, findings for general luxury confirm the findings for luxury food. Materialism and luxury are discussed interrelated (Hudders & Pandelaere, 2012; Wiedmann, Hennigs, & Siebels, 2007, 2009; Wong & Ahuvia, 1998). However, the mean values of evaluation in this factor imply that the materialistic content is rejected by the majority for luxury food. Two reasons can be envisaged: either the consumption of luxury food is generally less based on materialistic motives, or the findings just reflect the postulated shift away from an purely material definition of luxury.

The fifth factor displays the value of *Usability* as defined by Wiedmann, Hennigs, and Siebels (2007, 2009). Accordingly, usability is interpreted as either based on product properties or on consumer needs. All items included for this value in the questionnaire have loadings with absolute values above 0.5. The highest is on the aspect of uselessness (-0.849) which refers to consumers' needs. As with Wiedmann, Hennigs, and Siebels (2009), the loadings for both negative items on luxury food have negative loadings while the item of luxury foods being pleasant loads positively. This finding basically expresses a benevolent attitude toward luxury food. The association of luxury being useless (e.g. Dubois, Czellar, & Laurent, 2005) cannot be confirmed in terms of luxury food. An explanation could be that luxury is at least indirectly interrelated to health. Participants were asked to evaluate to what extent they associate 24 given food product attributes with luxury (five-point Likert scale with -2=full disagreement to 2=full agreement). High quality was in mean evaluated with 0.57. Quality of food products, e.g. in terms of organic food, in turn contributes to higher health values (Hoffmann, Wolf, & Staller, 2007).

The attribute of *Uniqueness* is displayed in the sixth factor, which again closely corresponds to the findings for general luxury brands (Vigneron & Johnson, 2004; Wiedmann, Hennigs, & Siebels, 2007, 2009). The loadings of all three items are in the high range while the item "True luxury food cannot be mass-produced." has the highest (0.793) as well as the highest positive mean value (0.81). The unsuitability of mass-production for luxury food is thus identified to be a crucial characteristic. This is not only congruent with research results on general luxury but it also reminds of the negative attitude toward agricultural mass production or intensive livestock farming recently found for German consumers (Dierks, 2007; Busch, Kayser, & Spiller, 2013).

Finally, the seventh factor addresses the financial dimension in terms of food *price*. The formulation of items does not relate directly toward luxury food but toward food quality and the significance of prices for purchase decisions. This is based on literature finding that price has a moderating function in defining general luxury, and for example, influences the perceived quality of a luxury product (Parguel, Delécolle, & Valette-Florence, 2014; Wiedmann, Hennigs, & Siebels, 2007, 2009). In terms of food, the results show that neither the signaling function of price for quality can be confirmed. Nor do they provide evidence that prices either provoke purchase intention or are indecisive. All mean values are slightly negative. A correlation analysis with Kendall-Tau-b and Spearman-Rho revealed that the price items correlate significantly but very weakly with income. The Spearman-Rho test value for income and the statement "When buying food, the price is not decisively." is the highest with 0.165. Thus, other factors must exist that influence the attitude toward food prices. Following on the postulated recent trend in Germany toward less price sensitivity of consumers in the food market (e.g. Institut für Demoskopie Allensbach, 2014), the findings on mean values of close to zero could mirror uncertainty and highly contrary attitudes in the early phase of those motivational changes.

Explorative factor analysis of preferred luxury food items

The second explorative factor analysis, structuring 21 possible luxury food-related items by means of consumption frequency, reveals three stable factors (Table 4). The lowest Cronbach's alpha value is 0.741, meaning that they are all considered as satisfactory (Bland & Altman, 1997). The loadings of all items are above 0.5, thus no item is sorted out. The first factor *Uniqueness, Price and Prestige* summarizes products that correspond to the financial, functional and social dimension of the conceptual model. The second factor *Sustainability and Authenticity* represents the dimension of New Luxury while the third factor includes *food items with a high value of self-identity*. It reflects the individual dimension of luxury.

Table 4. Explorative factor analysis of preferred luxury food items[1]

Factor 1: Uniqueness, Price and Prestige (Cronbach's alpha: 0.926, explained share of overall variance: 30.876%)	Factor loadings	μ ; σ
Dish in a fine-dining restaurant	0.821	-1.10; 0.994
A private wine cellar with a selection of expensive wines	0.819	-1.41; 0.992
Excellent vintage Bordeaux wine	0.787	-1.25; 1.012
Oyster, Caviar	0.756	-1.49; 0.845
Champagne	0.750	-1.23; 0.967
Meal in an insider restaurant	0.748	-1.02; 1.030
Twelve-year-old balsamic vinegar	0.725	-1.24; 1.049
A piece of meat from a particular type of farming (e.g. Kobe beef from Japan)	0.701	-1.27; 0.953
High-priced mineral water	0.683	-1.26; 1.072
Specialties from abroad (e.g. original Parmesan cheese)	0.602	-0.47; 1.175
Exclusive kitchen appliances (e.g. steamer, Teppanyaki grill)	0.597	-1.15; 1.084
Wine of a small, upcoming wine grower	0.577	-0.95; 1.139
Factor 2: Sustainable and authentic food items (Cronbach's alpha: 0.808, explained share of overall variance: 14.714%)	**Factor loadings**	μ ; σ
Organic food products	0.776	-0.12; 1.204
Regional products (e.g. milk of the nearby farm house)	0.774	0.08; 1.264
Fresh fruits and vegetables from the local farmers' market	0.709	0.41; 1.234
Fair trade coffee	0.589	-0.76; 1.248
Hand-made food products (e.g. marmalade, cakes)	0.556	0.51; 1.138
Factor 3: Food items with a high value of self-identity (Cronbach's alpha: 0.741, explained share of overall variance: 12.139%)	**Factor loadings**	μ ; σ
My favorite dish self-made	0.786	0.91; 1.009
A dinner which I enjoy in a state of tranquility after long working hours	0.735	0.73; 1.022
A piece of hand-made cake, eaten with family/friends	0.728	0.77; 1.024
Food items which have a special meaning to me (e.g. childhood memories, favorite dishes of relatives/a good friend)	0.666	0.09; 1.025

[1]Food items evaluated on a five-point Likert scale (from 2=frequent consumption to -2=no consumption) based on the question "How often do you consume the following food items?"; factor analysis on the basis of varimax rotation with Kaiser normalization; 5 iterations, Kayser-Meyer-Olkin criterion: 0.941; Bartlett significance: 0.000; cumulated explained share of overall significance: 57.729%.

Three-stage cluster analysis

The cluster analysis is formed by seven luxury food dimensions, which were revealed in the first factor analysis. It was conducted as follows: first, outliers identified with the Single-Linkage method were eliminated; next, the optimal number of clusters was determined based on the Ward method and finally, the end partitions were estimated with the K-means cluster analysis (Hair, Tatham, Anderson, & Black, 1998). Four consumer clusters were identified, for which different perceived dimensions of luxury food are relevant (Table 6). Post Hoc tests revealed that the luxury food dimensions *Sustainability and Authenticity* (factor 3), *Materialism* (factor 4), *Uniqueness* (factor 6) and *Price* (factor 7) primarily contribute to the demarcation of clusters while *Prestige and Hedonism* (factor 1) discriminates only cluster 1 and 2 significantly. The discriminant analysis (Hair et al., 1998) shows that 97.8% of all cases were classified correctly into clusters 1 to 4. Significant differences among the cluster formatting variables were revealed. The usefulness of the segmentation is thus given (Table 5).

Table 5. Discriminant analysis of cluster forming variables[1]

Discriminant Function	Eigenvalue	Canonial Correlation	Wilk's Lambda	Chi-Quadrat	Sig.
1	1.790	0.801	0.078	1873.327	0.000
2	1.421	0.766	0.216	1121.785	0.000
3	0.911	0.690	0.523	474.220	0.000
			Function 1	**Function 2**	**Function 3**
Centroids (group means)					
Cluster 1			-1.185	-0.721	-1.037
Cluster 2			1.971	-0.752	0.063
Cluster 3			0.341	2.530	-0.452
Cluster 4			-1.014	-0.038	1.479
Significant variables (structure matrix)					
Factor 1: Prestige and Hedonism			0.307	-0.190	0.111
Factor 2: Quality and Self-Identity			-0.541	0.265	-0.042
Factor 3: Sustainability and Authenticity			0.012	-0.127	-0.646
Factor 4: Materialism			0.453	0.763	-0.266
Factor 5: Usability			-0.023	0.198	0.187
Factor 6: Uniqueness			0.046	0.084	0.418
Factor 7: Price			0.095	-0.177	0.051

[1] Ninety-seven and eight-tenths percent of the cases were classified correctly.

Cluster 1: The Sustainability- and Health-oriented Realists

The first cluster contains 29.9% (n=221) of the participants. The proportion of females is 56.6% and the mean age is 50.6 years. The cluster is characterized by a high amount of people with lower education and middle net household incomes. A share of 48.3% of respondents report that luxury food is not important or not important at all for them. With respect to the willingness to pay for luxury food, most of the sustainability-oriented and health-oriented realists are indifferent (43%), and 33.6% are willing or very willing. Further, the members of the clusters show a clear focus on healthiness. In comparison to the other clusters, the new luxury dimension *Sustainability and Authenticity* has the highest mean value. In contrast, the aspects of *Prestige and Hedonism*, *Materialism* and *Price*, which are also associated with a more traditional understanding of luxury, show the second lowest cluster centers among all clusters. The target group also clearly rejects the facets of *Usability* and *Uniqueness*, which is reflected by the lowest mean values in cluster comparison.

In summary, the sustainability- and health-oriented realists are characterized by the hypothesized modern understanding of luxury, manifested in the clear sustainability- and authenticity-orientation. Additionally, the cluster has a focus on *Quality and Self-identity*. Except for the quality-facet, the target group rejects all the other typical established luxury values. The dominating buying motives of these quality-oriented consumers are *Sustainability and Authenticity*. Therefore, this target group does not prefer food products that are characterized by classical luxury attributes.

Cluster 2: The Prestige-seeking Connoisseur

This cluster contains 27.9% (n=206) of the participants. A share of 43.7% is female and the mean age is 43.7 years, the lowest of all clusters. The highest share is employees in the middle income classes. Concerning the perceived importance of luxury food, this cluster shows no clear profile. Most people are indifferent. More than one-third is willing to pay higher prices for food with luxury value. The most-often-mentioned categories where people are most likely to spend money on luxury food are meat and meat products, fish and fish products as well as fruits and vegetables. The factor *Uniqueness, Prestige, Price* from the second factor analysis, which summarizes expensive luxury food items, has the highest positive cluster center in this cluster among all others. Food items mirroring self-identity are instead less consumed, the mean value of the corresponding factor is the lowest in this cluster.

Organic or regional food, represented by factor 2 of the second factor analysis, is consumed slightly more than the average among all clusters. The results for the perceived dimensions of luxury food (cluster-forming factors) show a comparatively high agreement for *Prestige and Hedonism* and the lowest agreement for the dimension of *Self-identity and Quality*. The respondents agree more with the factor price, which reflects a positive attitude toward *price* as an indicator for good quality, in this cluster than in others. In summary, a group of consumers is found where the classical luxury dimensions of price and prestige, combined with hedonistic motivators like self-realization and cultural development, are perceived to determinate luxury food value. The social content of luxury food consumption is valued the highest in this cluster, while items expressing introversion and self-identity have the lowest means. Generally, in comparison to the other identified consumer groups, this cluster can be categorized as having an affinity toward classical luxury food items, which are defined by high prices and prestigious brands. The level of consumption of those luxury foods is often less than desired by consumers of this cluster. Materialistic statements like "It sometimes bothers me quite a bit that I can't afford to buy all the luxury food products I'd like." are agreed with more frequently than in other segments.

Table 6. Segmentation of consumers based on perceived dimensions of luxury food

Cluster formation	Cluster center (SD in brackets)				F	Sig.	Games Howell Post Hoc Test[1]
Dimensions of Luxury Food	Cluster 1 (N=221=29.9%)	Cluster 2 (N=206=27.9%)	Cluster 3 (N=127=17.2%)	Cluster 4 (N=185=25.0%)			
Factor 1: Prestige and Hedonism	-0.309 (0.724)	0.682 (1.110)	-0.384 (0.895)	-0.126 (0.843)	56.607	0.000	Cluster 1,2; 2,3; 2,4
Factor 2: Quality and Self-Identity	0.388 (0.619)	-0.997 (1.034)	0.396 (0.679)	0.374 (0.711)	153.064	0.000	Cluster 1,2 2,3; 2,4
Factor 3: Sustainability and Authenticity	0.632 (0.762)	0.067 (0.772)	-0.021 (0.958)	-0.815 (0.933)	98.889	0.000	Cluster 1,2; 1,3; 1,4; 2,4; 3,4
Factor 4: Materialism	-0.541 (0.609)	0.202 (0.757)	1.470 (0.767)	-0.588 (0.539)	308.708	0.000	Cluster 1,2; 1,3; 2,3; 2,4; 3,4
Factor 5: Usability	-0.297 (0.898)	-0.175 (0.823)	0.392 (1.118)	0.280 (1.055)	21.603	0.000	Cluster 1,3; 1,4; 2,3; 2,4
Factor 6: Uniqueness	-0.508 (0.996)	0.050 (0.928)	0.037 (0.912)	0.525 (0.841)	42.308	0.000	Cluster 1,2; 1,3; 1,4; 2,4; 3,4
Factor 7: Price	-0.037 (0.925)	0.315 (0.962)	-0.427 (1.000)	-0.014 (1.017)	15.490	0.000	Cluster 1,2; 1,3; 2,3; 2,4; 3,4

Cluster description	Cluster 1 (N=221=29.9%)		Cluster 2 (N=206=27.9%)		Cluster 3 (N=127=17.2%)		Cluster4 (N=185=25.0%)		Pearson-Chi-Quadrat	Asymp. Sig. (bilateral)	F	Sig.
Luxury food purchasing behavior (cross classification with second explorative factor analysis)	**Cluster center (SD in brackets)**								2130.446	0.333		
Factor 1: Uniqueness, Price and Prestige	-0.259 (0.710)		0.557 (1.202)		-0.052 (0.950)		0.164 (1.027)				24.768	0.000
Factor 2: Sustainability and Authenticity	0.401 (1.012)		0.051 (0.821)		-0.204 (0.931)		-0.198 (1.018)				16.818	0.000
Factor 3: Self-Identity	0.167 (0.944)		-0.249 (0.995)		0.096 (0.973)		0.102 (1.023)				7.107	0.000
	n	**%**	**n**	**%**	**n**	**%**	**n**	**%**	Pearson-Chi-Quadrat	Asymp. Sig. (bilateral)		
Perceived importance of Luxury Food[2]									36.011	0.000		
Very important or important	37	16.8	57	27.8	41	32.3	31	16.8				
Indifferent	77	35.0	91	44.4	44	34.6	66	35.9				
Not important or not important at all	106	48.2	57	27.8	42	33.0	87	47.3				

Willingness to pay for Luxury Food[3]									30.824	0.002	
Very willing or willing	74	33.5	79	38.6	64	50.4	81	43.8			
Indifferent	95	43.0	74	36.1	45	35.4	56	30.3			
Not willing or not willing at all	52	23.5	62	25.4	18	14.2	48	25.9			
Categories of high willingness to pay for luxury foods[4]											
Bread	95	56.2	63	40.9	54	50.0	51	37.2	29.972	0.003	
Chocolate	54	32.1	69	45.1	47	43.1	67	49.3	28.671	0.004	
Meet, meet products	119	71.7	94	61.4	75	70.1	95	70.3	16.354	0.176	
Fish, fish products	109	65.7	84	54.9	72	67.3	86	63.2	15.704	0.205	
Wine, sparkling wine, champagne	51	30.9	69	44.8	42	39.3	67	49.7	27.873	0.006	
Mineral water, limonades, juices	29	17.3	40	25.9	25	23.1	22	16.0	21.114	0.049	
Fruits, vegetables	112	66.2	78	50.7	64	58.7	72	52.5	27.635	0.006	
Convenience food	10	6.1	31	20.3	13	12.1	10	7.6	39.541	0.000	
Spices	71	42.0	58	37.6	50	45.8	66	48.2	27.445	0.007	
Cheese	86	51.2	77	50.0	66	61.7	76	56.3	18.750	0.095	
Fats, oil	59	34.9	49	31.8	45	41.3	53	38.7	22.404	0.033	
Yoghurt	57	34.1	52	33.8	33	30.8	37	27.0	28.869	0.004	

Preferred retailer[5]											
Discounter	175	79.2	174	84.5	116	91.3	140	75.7	28.774	0.004	
Local farmers' market	57	25.8	62	30.1	9	7.1	40	21.6	32.043	0.001	
Whole food market	28	12.7	38	18.4	9	7.1	11	5.9	40.585	0.000	
Farm shop	21	9.5	37	18.0	7	5.5	14	7.5	30.098	0.003	
Deli shop	16	7.2	32	15.6	7	5.5	15	8.1	29.209	0.004	
Specialized shop (e.g. butcher)	82	37.1	85	41.2	29	22.8	61	32.9	22.235	0.035	
Special diet styles[6]											
Vegetarian	11	5.1	20	9.8	3	2.4	4	2.2	48.713	0.000	
Vegan	3	1.4	12	5.9	4	3.2	2	1.1	62.132	0.000	
Focus on healthiness	122	55.2	101	49.1	61	48.1	85	45.9	18.627	0.000	
Focus on lankiness	38	17.2.	40	19.4	20	15.7	23	12.5	25.424	0.013	
Eating in small quantities	62	28	60	29.1	32	25.4	39	21.1	20.316	0.061	

Socio-demographic data	n	%	n	%	n	%	n	%	Pearson-Chi-Quadrat	Asymp. Sig. (bilateral)
Gender									12.708	0.005
Male	96	43.4	116	56.3	55	43.3	105	56.8		
Female	125	56.6	90	43.7	72	56.7	80	43.2		
Age[7]									57.763	0.000
18–29 years	20	9.0	46	22.4	31	24.4	22	11.9		
30–39 years	26	11.8	51	24.9	20	15.7	17	9.2		
40–49 years	54	24.4	34	16.6	28	22.0	43	23.2		
50–59 years	55	24.9	37	18.0	21	16.5	54	29.2		
60–69 years	49	22.2	28	13.7	17	13.4	34	18.4		
70–75 years	17	7.7	9	4.4	10	7.9	15	8.1		
Region of origin[8]									8.855	0.451
North	47	21.3	37	18.0	19	15.0	33	17.8		
East	27	12.2	44	21.4	23	18.1	31	16.8		
South	63	28.5	54	26.2	37	29.1	58	31.4		
West	84	38.0	71	34.5	48	37.8	36	34.1		
Marital status										
Single	39	17.6	50	24.3	32	25.2	41	22.2	36.163	0.021
Single with child(ren)	10	4.5	4	1.9	7	5.5	2	1.1		
In a relationship	35	15.8	31	15.0	26	20.5	27	14.6		
In a relationship with child(ren)	8	3.6	10	4.9	9	7.1	9	4.9		
Married	60	27.1	59	28.6	26	20.5	56	30.3		

Married with child(ren)	52	23.5	40	19.4	22	17.3	48	25.9		
Widowed	16	7.2	8	3.9	3	2.4	1	0.5		
No comment	1	0.5	4	1.9	2	1.6	1	0.5		
School education[9,10]									33.063	0.103
Lower education	89	40.3	71	34.5	50	39.4	56	30.3		
Middle education	64	29.0	81	39.4	48	37.9	62	33.5		
Higher education	63	28.5	48	23.3	26	20.5	64	34.6		
No comment	5	2.3	6	2.9	3	2.4	3	1.6		
Professional education[10]									34.091	0.277
Still in education	4	1.9	8	3.9	5	3.9	0	0.0		
Apprenticeship/Master craftman	127	57.4	108	52.4	76	59.8	103	55.7		
University/University of Applied Sciences	59	26.7	59	28.6	24	18.6	60	32.4		
No comment	31	14	31	15.1	22	17.3	22	12		
Occupation[10]									66.510	0.020
Student/Trainee	5	2.3	6	3.0	5	3.9	4	2.2		
Employee	90	40.8	91	44.1	46	36.2	76	41.1		
Self-employer	17	7.7	13	6.3	11	8.7	25	13.5		
Academic freelancer	4	1.8	2	1.0	1	0.8	5	2.7		
Official	9	4.1	12	5.8	2	1.6	7	3.8		
Worker	30	13.6	34	16.5	22	17.3	17	9.2		
Retired	49	22.2	29	14.1	26	20.5	39	21.1		
No comment	5	2.3	4	2.0	4	3.1	5	2.7		

Net household income									80.150	0.000		
Less than 900 Euro	11	5.0	14	6.8	18	14.2	9	4.9				
900–less than 1300 Euro	26	11.8	21	10.2	19	15.0	12	6.5				
1300–less than 1500 Euro	14	6.3	7	3.4	12	9.4	10	5.4				
1500–less than 2000 Euro	31	14.0	38	18.4	20	15.7	12	6.5				
2000–less than 2600 Euro	32	14.5	27	13.1	13	10.2	29	15.7				
2600–less than 3600 Euro	28	12.7	32	15.5	16	12.6	22	11.9				
3600–less than 5000 Euro	16	7.2	21	10.2	11	8.7	33	17.8				
5000–less than 7600 Euro	14	6.3	17	8.3	5	3.9	15	8.1				
7600–less than 9000 Euro	10	4.5	4	1.9	3	2.4	10	5.4				
9000–less than 12.600Euro	6	2.7	2	1.0	1	0.8	5	2.7				
12.600–less than 18.000 Euro	1	0.5	1	0.5	0	0.0	6	3.2				
More than 18.000 Euro	0	0.0	4	1.9	0	0.0	2	1.1				
No comment	32	14.5	18	8.7	9	7.1	20	10.8				

[1]Mean value differences among the clusters which are significant on the level 0.05.

[2]Question "How important do you find luxury on food?", evaluated on a five-point Likert scale (from 2=very important to -2=not important at all).

[3]Question "Are you willing to pay a higher price for foods that you connect with luxury?", evaluated on a five-point Likert scale (from 2=very willing to -2=not willing at all).

[4]Question "In which of the following food categories are you willing to pay a higher price for food items you perceive to be a piece of personal luxury?", 12 categories were to evaluate on a five-point Likert scale (from 2=very willing to -2=not willing at all), values are named for "Willing" or "Very willing".

[5]Question "Where do you predominantly buy your foods?", answering options are evaluated on a five-point Likert scale (from 2=very often to -2=never), percentages are given for "Often" or "Very often".

[6]Question "Which diet style do you cultivate?", answering options are evaluated on a five-point Likert scale (from 2=applies completely to -2=doesn´t apply at all), values are named for "Applies" or "Applies completely".

[7]Mean age: Cluster 1: 50.6 years; Cluster 2: 43.7 years; Cluster 3: 44.9 years; Cluster 4: 49.9 years.

[8]North: Schleswig-Holstein, Hamburg, Mecklenburg-Vorpommern, Lower Saxony, Bremen; East: Berlin, Brandenburg, Saxony-Anhalt, Saxony, Thuringia; South: Bavaria, Baden-Wuerttemberg; West: North Rhine-Westphalia, Hesse, Rhineland-Palatinate, Saarland.

[9]Lower education: no school leaving certificate/primary school/lower secondary school; Middle education: secondary school, polytechnic high school, technical school; Higher education: general or specialized higher education.

[10]Data belong to the main earner in the particular household.

Cluster 3: The Wishful Materialists

The smallest cluster consists of 127 participants (17.2%) with 56.7% females and a mean age of 44.9 years. The group is characterized by a high share of people with low and middle education. Concerning the perceived importance of luxury food the cluster does not show a clear profile. Instead, there is a clear ambivalence (very important or important: 32.3 %; indifferent: 34.6%; not important or not important at all: 33.0%). Although the cluster contains a high share of people with low to middle net household incomes, the amount of cluster members with a willingness to pay for luxury food (50.4%) is noticeable. In comparison to the other clusters, the materialistic value of food products is most important, the mean value for *Materialism* (1.47) is the highest among all mean values in the cluster analysis. Further, the typical luxury facets, *Quality and Self-identity* as well as *Usability*, also have highest cluster centers in comparison to the other clusters. In contrast, high prices of food products are of lowest importance for purchase decision among all clusters. Concerning the dimensions *Uniqueness* as well as *Sustainability and Authenticity*, the group of the wishful materialists has the second lowest mean values. In summary, this cluster is characterized by a high materialism orientation as well as a clear focus on *Quality and Self-identity* and *Usability*. However, high prices are rejected and *Uniqueness* is less important. The factor *Prestige and Hedonism* is evaluated negatively. It can be assumed that this segment has the need for luxury food products coupled with a limited budget.

Cluster 4: The Introvert Specialties-seeker

This cluster contains 25% (n=185) of all participants and is slightly male-dominated, with a mean age of 49.9 years. The share of people with high education and high net household income is comparatively high. Forty-three and eight-tenths percent are very willing or willing to pay for luxury food value, even though most people stated to not find luxury food value important. The preferred categories for luxury food are again meat and meat products, fish and fish products, and cheese. Special food styles are less distinctive compared to other clusters. Instead, 68.1% fully agreed or agreed to the statement "I eat a bit of everything." The focus on healthiness is less pronounced than in other clusters but still contains 45.9%. In the analysis of luxury food purchasing behavior, this cluster shows the second highest preference for unique, expensive and prestigious food items. They are most likely to spend money on specialties from abroad, meals in insider- or fine dining restaurants or wine of a small, upcoming wine grower. At the same time, food products that are associated with *Sustainability and Authenticity* are comparatively less consumed. The mean value for food products reflecting *Self-identity* is the second highest among all clusters. In correspondence, the decisive dimensions of luxury food are *Uniqueness, Self-identity and Quality*, and *Usability*, in descending order of importance. This target group does not use prices to determine the perceived luxury value of foods. The group further tends to reject *Sustainability and Authenticity* as a luxury dimension. The same applies for *Prestige, Hedonism* and *Materialism*. The refusal of *Materialism* is the most notable. The factor's mean value in this cluster is the lowest among all others. The statements "My life would be better if I bought certain luxury food products I don't consume." for example, is evaluated by a mean value of -1.48, reflecting almost complete disagreement.

Discussion and next research steps

Discussion

Dimensions of luxury food value strongly correspond to the perceived dimensions of luxury in general (Wiedmann, Hennnigs, & Siebels, 2009). As in the definition of a general luxury good, luxury food is associated with a social, individual, functional and financial value. What is shown to be different is the intersection of the individual dimension with two others. In one factor, hedonism is combined with a social dimension, in terms of prestige (factor 1). In an additional one, self-identity is conjoined with a preference for special food qualities (factor 2). Overall, the individual dimension is found to define luxury food value in terms of three specifications; the third is materialism. The social aspect of luxury food is concentrated on prestige, while functionality matters in terms of usability, quality and uniqueness. The revealed seventh factor of price defines the financial dimension.

In addition to the perceived definitional dimensions of general luxury, the analysis contains investigations into new luxury values. It was tested to what extent sustainability and authenticity play a role for the definition of luxury food. In the factor analysis, one separate factor was revealed, comprising the aspects of organic food, environment-friendliness, fairness in food production and origin.

Based on those factors, a cluster analysis was used to estimate four consumer segments in which the definitional dimensions have different significant meanings. It was found that two segments are target groups for the premium food market, which is associated with high quality and high prices (cluster 2 and 4). One further segment is formed by the target group for foods with high values of sustainability and authenticity (cluster 1) and an additional group is highly materialistic oriented but is not expected to buy at premium food markets due to its comparatively low income (cluster 3). At the same time, this is expected to be highly affine to marketing strategies that are geared towards functional luxury values and self-identity for food products of low or average price ranges.

Consistently across all segments most people are willing to spend money on luxury food. The categories where participants are most likely to spend money for luxury food are meat (products), fish (products), fruits and vegetables as well as cheese. Other studies show that those categories, especially meat and fish, are also crucial in transparency, animal welfare and general quality and health concerns of consumers (Issanchou, 1996; Verbeke, Vermeir, & Brunsø, 2007; Verbeke & Viaene, 2000). The need for transparency and quality of consumers clearly differs depending on product categories of food, whereas "sensitive food products" like meat, meat products, eggs and fish are associated with the highest requirements of information about the product (SGS, 2014). These food products partially were involved in food scandals (e.g. dioxin in eggs, rotten meat scandal), which suggests the assumption that consumers have an increased need for transparency with respect to quality.

In summary, neither the above-mentioned first nor second research hypothesis can be rejected. On the one hand, the conceptual model, which general luxury research inspires (Wiedmann, Hennigs, & Siebels, 2007, 2009), reveals stable and plausible factors, providing a base for a reasonable consumer segmentation. On the other hand, new luxury values, represented by sustainability and authenticity in this case, are revealed to be another dimension in the definition of luxury foods. Even though, the cluster analysis shows that this factor's means are low or negative among segments which are potential target groups for luxury foods in a classical sense of expensive, unique premium products (clusters 2 and 4). Segment one instead, which is the least affine toward a traditional, materialistic understanding of luxury, has the highest rating for the factor of sustainability and authenticity. This implicates that buyers of organic and Fair Trade food products have to be differentiated from buyers of expensive premium food brands. The latter rather associate luxury foods with more classical dimensions, especially prestige and hedonism (cluster 2) and uniqueness (cluster 4). Among them, the attitude towards sustainability and authenticity is close to indifference (cluster 2) or even negative (cluster 4). The first group instead, is indeed willing to pay higher prices for sustainable, authentic and healthy food products but they clearly reject the connection to luxury in the sense of any traditional dimension. According to Kapferer (2010), an averse attitude towards luxury is based on the association of luxury as an instrument to create inequality within the society and to force the demonstration of wealth and class. If the individual perception of luxury is mainly based on traditional, externally-oriented dimensions, reluctance toward prestige and conspicuous consumption will cause that sustainable and authentic attributes of luxury food will be denied. Padilla Bravo et al. (2013) found another possible reason. They consider that marketing strategies subsumed under "exclusiveness" could threaten organic food marketing. Consumers might associate exclusive gourmet and specialties with conventional production methods. This would delete the link between sustainable, authentic foods and luxury as far as gourmet and specialties are defined as luxury food. The marketing thus should take the different attitudes toward luxury in both target groups into consideration. Premium food buyers will rather be willing to pay a higher price when a particular product is marketed via the whole catalogue of dimensions except sustainability and authenticity. Consumers from the first segment can be reached by strategies based on sustainability, authenticity and health.

What cannot be clearly differentiated between those two categories of consumers is the dimension of quality and self-identity. This is only rejected by cluster two, all other means are close to 0.4. It thus can be concluded that emphasizing these two product attributes in marketing strategies is applicable for addressing more than 70% of consumers.

Next research steps

This paper empirically investigates the definitional dimensions of luxury food and further defines consumer segments in which the revealed dimensions have varying significance. It can serve as a base for target group-oriented marketing strategies on food markets, especially for premium and high quality food products. Further, the results contribute to a better understanding of luxury consumption in general. In the analysis, empirical evidence was found for the recent significance of new luxury values in terms of sustainability and authenticity. The conceptual model appears to be an appropriate instrument for the measurement of perceptions on luxury and is easily adaptable to new luxury consumption patterns. In this way, the research in luxury marketing has delivered an extended approach to empirically investigate various luxury markets. For the luxury food market, this is the first study of its kind. Nevertheless, these results need to be validated in further research, e.g. by means of a path least square structural equation model (Henseler, Ringle, & Sinkovics, 2009).

Further tests should focus on how the revealed factors actually influence the latent luxury food dimensions and finally contribute to the perceived luxury value of a food. Accordingly, Wiedmann, Hennigs, and Siebels (2009) propose a confirmatory factor analysis to determine the relationships of variables in their conceptual model. Hartmann et al. (2013) provide an example for a path least square regression analysis on consumer attitudes.

Against the background that different new consumption motives on luxury markets are postulated, it would also be useful to expand the model accordingly. Depending on what market is investigated, new (internally-oriented) dimensions of luxury could be integrated. This would help to fill the recent vacuum of empirical studies that analyze the described change in luxury consumption patterns on different luxury markets.

It might be of further interest to investigate luxury food consumption empirically with an international database. Both the consumption of luxury goods and food consumptions patterns differ among cultures and nations (Gracia & Albisu, 2001; Wong & Ahuvia, 1998). It thus can be assumed that the perceptions on luxury food will be different in other countries. In a globalized world of international marketing strategies, this might give important implications for targeting consumer segments cross-culturally.

References

Albrecht, C.-M., Backhaus, C., Gurzki, H., & Woisetschläger, D. M. (2013). Value Creation for Luxury Brands through Brand Extensions: An Investigation of Forward and Reciprocal Effects. *Marketing ZFP – Journal of Research and Management, 35*(2), 91-103.

Allérès, D. (1993). L'univers du luxe. *Regards sur l'actualité, 187,* 3-26.

Appadurai, A. (Ed.) (1988). Introduction: Commodities and the Politics of Value. *The Social Life of Things in Cultural Perspectives,* 3-63.

Ascheberg, C., Meurer, J., & Oesterling, A. (2012). The Luxury Universe–Angebots-und Kundensegmentierung globaler Luxusmärkte als Basis für erfolgreiche Positionierungs-strategien. In C. Burmann, V. König, & J. Meurer (Hrsg.), *Identitätsbasierte Luxusmarken-führung* (S. 85-101). Springer Fachmedien Wiesbaden.

Atwal, G., & Williams, A. (2009). Luxury brand marketing–the experience is everything!. *Journal of Brand Management, 16*(5), 338-346.

Baker, S., Thompson, K. E., Engelken, J., & Huntley, K. (2004). Mapping the values driving organic food choice: Germany vs the UK. *European Journal of Marketing, 38*(8), 995-1012.

Bellaiche, J., Kluz, M., Mei-Pochtler, A., & Wiederin, E. (2012). *Luxe Redux: raising the bar for selling of luxuries.* Boston Consulting Group, Boston.

Berry, C. J. (1994). *The Idea of Luxury: A Conceptual and Historical Investigation* (Vol. 30). Cambridge University Press.

Beverland, M. (2006). The 'real thing': branding authenticity in the luxury wine trade. *Journal of Business Research, 59*(2), 251-258.

Beverland, M. B. (2005). Crafting brand authenticity: the case of luxury wines. *Journal of Management Studies, 42*(5), 1003-1029.

Bilharz, M., & Belz, F. M. (2008). Öko als Luxus-Trend: Rosige Zeiten für die Vermarktung „grüner " Produkte?. *Marketing Review St. Gallen, 25*(4), 6-10.

Bland, J. M., & Altman, D. G. (1997). Statistics notes: Cronbach's alpha. *Bmj, 314*(7080), 572.

Blevis, E., Makice, K., Odom, W., Roedl, D., Beck, C., Blevis, S., & Ashok, A. (2007). Luxury & new luxury, quality & equality. In *Proceedings of the 2007 conference on Designing pleasurable products and interfaces* (pp. 296-311). ACM.

Brunsø, K., Grunert, K. G., & Bredahl, L. (1996). *An analysis of national and cross-national consumer segments using the food-related lifestyle instrument in Denmark, France, Germany and Great Britain.* Aarhus, Denmark: MAPP.

Brunsø, K., Scholderer, J., & Grunert, K. G. (2004). Testing relationships between values and food-related lifestyle: results from two European countries. *Appetite, 43*(2), 195-205.

Busch, G., Kayser, M., & Spiller, A. (2013). „Massentierhaltung "aus VerbraucherInnensicht– Assoziationen und Einstellungen. *Jahrbuch der Österreichischen Gesellschaft für Agrarökonomie, 22*(1), 61-70.

Dierks, L. H. (2007). Does trust influence consumer behavior? *German Journal of Agricultural Economics, 56*(2), 106-111.

Dietler, M. (1990). Driven by drink: the role of drinking in the political economy and the case of Early Iron Age France. *Journal of Anthropological Archaeology, 9*(4), 352-406.

Dietler, M. (1996). Feasts and commensal politics in the political economy: Food, power, and status in prehistoric Europe. In P. Wiessner & W. Schiefenhoevel (Eds.), *Food and the status quest: An interdisciplinary perspective* (pp. 87-125). Providence, RI: Berghahn.

Dubois, B., & Laurent, G. (1994). Attitudes toward the concept of luxury: An exploratory analysis. *Asia-Pacific Advances in Consumer Research, 1*(2), 273-278.

Dubois, B., & Paternault, C. (1995). Observations: Understanding the world of international luxury brands: The" dream formula". *Journal of Advertising Research, 35*(4), 69-76.

Dubois, B., Czellar, S., & Laurent, G. (2005). Consumer segments based on attitudes toward luxury: empirical evidence from twenty countries. *Marketing letters, 16*(2), 115-128.

Erickson, G. M., & Johansson, J. K. (1985). The role of price in multi-attribute product evaluations. *Journal of Consumer Research, 12*(2), 195-199.

Fabrigar, L. R., Wegener, D. T., MacCallum, R. C., & Strahan, E. J. (1999). Evaluating the use of exploratory factor analysis in psychological research. *Psychological methods, 4*(3), 272.

Fassnacht, M., Kluge, P. N. & Mohr, H. (2013). Pricing Luxury Brands: Specificities, Conceptualization and Performance Impact. *Marketing ZFP – Journal of Research and Management, 35*(2), 104-117.

Ford, J. K., MacCallum, R. C., & Tait, M. (1986). The application of exploratory factor analysis in applied psychology: A critical review and analysis. *Personnel Psychology, 39*(2), 291-314.

Glare, P. G. (1982). *Oxford latin dictionary*. Clarendon Press. Oxford University Press.

Gosden, C., & Hather, J. G. (Eds.) (2004). *The Prehistory of food: appetites for change*. Routledge.

Gracia, A. & Albisu, L. M. (2001). Food consumption in the European Union: Main determinants and country effects. Agribusiness, *17*(4), 469-488.

Grimm, P. (2010). *Social desirability bias*. Wiley International Encyclopedia of Marketing.

Gumerman Iv, G. (1997). Food and complex societies. *Journal of Archaeological Method and Theory, 4*(2), 105-139.

Hair, J. F., Tatham, R. L., Anderson, R. E., & Black, W. (2006). *Multivariate data analysis* (Vol. 6). Upper Saddle River, NJ: Pearson Prentice Hall.

Han, Y. J., Nunes, J. C., & Drèze, X. (2010). Signaling status with luxury goods: the role of brand prominence. *Journal of Marketing, 74*(4), 15-30.

Harper, G. C., & Makatouni, A. (2002). Consumer perception of organic food production and farm animal welfare. *British Food Journal, 104*(3/4/5), 287-299.

Hartmann, L., Kerssenfischer, F., Fritsch, T., & Nguyen, T. (2013). User Acceptance of Customer Self-Service Portals. *Journal of Economics, Business and Management, 1*(2), 150-155.

Hayden, B. (2003). Were luxury foods the first domesticates? Ethnoarchaeological perspectives from Southeast Asia. *World Archaeology, 34*(3), 458-469.

Henseler, J., Ringle, C. M., & Sinkovics, R. R. (2009). The use of partial least squares path modeling in international marketing. In R. R. Sinkovics & P. N. Ghauri (Eds.), *New Challenges to International Marketing (Advances in International Marketing, Volume 20)* (pp. 277-319). Emerald Group Publishing Limited.

Hoffmann, M., Wolf, G., & Staller, B. (2007). Lebensmittelqualität und Gesundheit. *Bio-Testmethoden und Produkte auf dem Prüfstand. Schwerin: Baerens & Fuss.*

Honkanen, P., Verplanken, B., & Olsen, S. O. (2006). Ethical values and motives driving organic food choice. *Journal of Consumer Behaviour, 5*(5), 420-430.

Hornig, T., Fischer, M., & Schollmeyer, T. (2013). The Role of Culture for Pricing Luxury Fashion Brand. *Marketing ZFP – Journal of Research and Management, 35*(2), 118-130.

Hudders, L., & Pandelaere, M. (2012). The silver lining of materialism: the impact of luxury consumption on subjective well-being. *Journal of Happiness Studies, 13*(3), 411-437.

Issanchou, S. (1996). Consumer expectations and perceptions of meat and meat product quality. *Meat Science, 43*(1), 5-19.

Jarvis, C. B., MacKenzie, S. B., & Podsakoff, P. M. (2003). A critical review of construct indicators and measurement model misspecification in marketing and consumer research. *Journal of Consumer Research, 30*(2), 199-218.

Joy, A., Sherry, J. F., Venkatesh, A., Wang, J., & Chan, R. (2012). Fast fashion, sustainability, and the ethical appeal of luxury brands. *Fashion Theory: The Journal of Dress, Body & Culture, 16*(3), 273-296.

Kapferer, J. N. (2010). All that glitters is not green: The challenge of sustainable luxury. *European Business Review*, 40-45.

Kemp, S. (1998). Perceiving luxury and necessity. *Journal of Economic Psychology, 19*(5), 591-606.

Kirig, A., & Ruetzler, M. H. (2007). Food-Styles. *Die wichtigsten Thesen, Trends und Typologien für die Genuss-Märkte.* Zukunftsinstitut-Studie. Kelkheim.

Kisabaka, L. (2001). *Marketing für Luxusprodukte* (Vol. 32). Dissertation, Fördergesellschaft Produkt-Marketing, Cologne.

Lueth, M., Spiller, A., Enneking, U. (2004): *Analyse des Kaufverhaltens von Selten- und Gelegenheitskäufern und ihrer Bestimmungsgründe für/gegen den Kauf von Öko-Produkten.* Projektabschlussbericht für das BMVEL im Rahmen des Bundesprogramms ökologischer Landbau, Goettingen.

Lupton, D. (1994). Food, memory and meaning: the symbolic and social nature of food events. *The Sociological Review, 42*(4), 664-685.

Magnusson, M. K., Arvola, A., Hursti, U. K. K., Åberg, L., & Sjoedén, P. O. (2003). Choice of organic foods is related to perceived consequences for human health and to environmentally friendly behaviour. *Appetite, 40*(2), 109-117.

Mason, R. (1993). Cross-cultural influences on the demand for status goods. *European Advances in Consumer Research*, 1, 46-51.

Meurer, J. (2012). Ebony or Ivory–wie glänzend ist die Zukunft des Luxus in Deutschland? Kritische Reflexionen zum Luxusmarkenmanagement. In C. Burmann, V. König, & J. Meurer (Hrsg.), *Identitätsbasierte Luxusmarkenführung* (S. 321-336). Springer Fachmedien Wiesbaden.

Meurer, J., & Manninger, K. (2012). Quo vadis globale Luxusmarkenführung–Status, Trends und Top-Themen für die CMO-Agenda. In C. Burmann, V. König, & J. Meurer (Hrsg.), *Identitätsbasierte Luxusmarkenführung* (S. 13-31). Springer Fachmedien Wiesbaden.

Mohr, E. (2014). *Ökonomie mit Geschmack: die postmoderne Macht des Konsums*. Murmann Verlag DE.

Mueller-Stewens, G. (2013). Das Geschäft mit Luxusgütern. Universität St. Gallen.

Nestlé Deutschland AG (2012) (Hrsg.). *Nestlé Studie 2012, Das is(s)t Qualität*, Auszüge aus der Nestlé Studie 2012.

Nitzko, S., & Spiller, A. (2014). Zielgruppenansätze in der Lebensmittelvermarktung. In M. Halfmann (Hrsg.), *Zielgruppen im Konsumentenmarketing* (S. 315-332). Springer Fachmedien Wiesbaden.

Nunnally, J. (1978). C.(1978). Psychometric theory. New York: McGraw-Hill.

Padilla Bravo, C., Cordts, A., Schulze, B., & Spiller, A. (2013). Assessing determinants of organic food consumption using data from the German National Nutrition Survey II. *Food Quality and Preference, 28*(1), 60-70.

Page, C. (2006). *Innovation in Gourmet and Specialty Food and Drinks: Market Evolution and NPD in Super-premium and Healthy Products*. Business Insights Limited.

Parguel, B., Delécolle, T., & Valette-Florence, P. (2014). The impact of price display on perceptions of luxury: a masstige perspective. <halshs-00948953>.

Pflanz, C. (2003). Faszination Luxus. In C. Pflanz (Hrsg.), *Faszination* (S. 90-97). Gabler Verlag.

Puntoni, S. (2001). Self-identity and purchase intention: An extension of the theory of planned behavior. *European Advances in Consumer Research*, 5, 130-134.

Richins, M., & Dawson, S. (1992). A consumer values orientation for materialism and its measurement: Scale development and validation. *Journal of Consumer Research, 19*(3), 303–316.

Robinson, J. P., Shaver, P. R., & Wrightsman, L. S. (1991). Criteria for scale selection and evaluation. *Measures of personality and social psychological attitudes, 1*(3), 1-16.

Schallehn, M. (2012). Identitätsbasierte Führung von Luxusmarken unter besonderer Berücksichtigung der Marken-Authentizität am Beispiel von Bugatti und Maybach. In C. Burmann, V. König, & J. Meurer (Hrsg.), *Identitätsbasierte Luxusmarkenführung* (S. 53-67). Springer Fachmedien Wiesbaden.

Schwartz, S. H. (1992). Universals in the content and structure of values: theoretical advances and empirical tests in 20 countries. In M. P. Zanna (Ed.), *Advances in experimental social psychology* (vol. 25, pp. 1–65). San Diego, CA: Academic Press.

Serraf, G. (1991). Le produit de luxe: somptuaire ou ostentatoire?. *Revue française du marketing*, (132), 7-16.

SGS (2014). Vertrauen und Skepsis: Was leitet die Deutschen beim Lebensmitteleinkauf? SGS-Verbraucherstudie 2014: Ergebnisse einer bevölkerungsrepräsentativen Befragung. Hamburg: SGS Germany GmbH.

Sidali, K. L., & Hemmerling, S. (2014). Developing an authenticity model of traditional food specialties: Does the self-concept of consumers matter?. *British Food Journal, 116*(11), 1692-1709.

Statistisches Bundesamt (2013). Zahlen & Fakten. URL: https://www.destatis.de/DE/ZahlenFakten/GesellschaftStaat/StaatGesellschaft.html;jsessio nid=C96AEA9A9BCDEF2DAA17DF50700EB5A6.cae1. Last accessed: 31 January 2015.

Stockebrand, N., & Spiller, A. (2008). Authentizität als Erfolgsfaktor im Regionalmarketing: Eine erste Skizze. *Ernährung, Lebensqualität-Wege regionaler Nachhaltigkeit*, 145-166.

Tsai, S. P. (2005). Impact of personal orientation on luxury-brand purchase value. *International Journal of Market Research, 47*(4), 429-454.

Van der Veen, M. (2003). When is food a luxury?. *World Archaeology, 34*(3), 405-427.

Veblen, T. (2007). *The theory of the leisure class*. Oxford University Press.

Verbeke, W. A., & Viaene, J. (2000). Ethical challenges for livestock production: Meeting consumer concerns about meat safety and animalwelfare. *Journal of Agricultural and Environmental Ethics, 12*(2), 141-151.

Verbeke, W., Vermeir, I., & Brunsø, K. (2007). Consumer evaluation of fish quality as basis for fish market segmentation. *Food Quality and Preference, 18*(4), 651-661.

Vigneron, F., & Johnson, L. W. (1999). A review and a conceptual framework of prestige-seeking consumer behavior. *Academy of Marketing Science Review, 1*(1), 1-15.

Vigneron, F., & Johnson, L. W. (2004). Measuring perceptions of brand luxury. *The Journal of Brand Management, 11*(6), 484-506.

Wheatley, J. J., & Chiu, J. S. (1977). The effects of price, store image, and product and respondent characteristics on perceptions of quality. *Journal of Marketing Research, 14*(2), 181-186.

Wiedmann, K. P., Hennigs, N., & Siebels, A. (2007). Measuring consumers' luxury value perception: a cross-cultural framework. *Academy of Marketing Science Review, 7*(7), 333-361.

Wiedmann, K. P., Hennigs, N., & Siebels, A. (2009). Value-based segmentation of luxury consumption behavior. *Psychology & Marketing, 26*(7), 625-651.

Wong, N. Y., & Ahuvia, A. C. (1998). Personal taste and family face: Luxury consumption in Confucian and Western societies. *Psychology and Marketing, 15*(5), 423-441.

Wycherley, A., McCarthy, M., & Cowan, C. (2008). Speciality food orientation of food related lifestyle (FRL) segments in Great Britain. *Food quality and preference, 19*(5), 498-510.

Yeoman, I., & McMahon-Beattie, U. (2006). Luxury markets and premium pricing. *Journal of Revenue and Pricing Management, 4*(4), 319-328.

II.3 The Significance of Definitional Dimensions of Luxury Food

Authors: **Laura Hartmann, Achim Spiller**

Georg-August-University of Goettingen

For submission.

Abstract

The perceived dimensions of luxury are complex. Empirical studies provide evidence for a social, individual, functional and financial layer of luxury perceptions. Furthermore, it is postulated that those dimensions are not stable over time. Instead, some authors postulate a recent trend toward entirely new luxury definitions. Accordingly, sustainability, authenticity, quality and personality-related motives, like self-fulfillment, seem to push old patterns of prestige and conspicuousness into the background. What has not been empirically investigated is the question whether and to what extent these various aspects significantly contribute to the luxury value of a good. By means of a partial least squares structural equation analysis, this study shows the effects of seven identified, exploratory value dimensions on a perceived luxury value for foods. Luxury food serves as a suitable setting to investigate both, old and new motives for luxury consumption. The results show that all factors contribute significantly to luxury value, with functional and individual luxury facets having the strongest effects.

Keywords

Luxury Foods, Luxury Dimensions, Significance, PLS Path Modeling

Introduction

Considerations on general luxury consumption in Western cultures

Recently, international luxury markets have been flourishing, and researchers predict a further upward trend (Bellaiche et al., 2014; D´arpizio & Levato, 2014). In 2012, the global luxury market volume was estimated at around 1.42 trillion Euros, and 14.9% of this was based on personal luxury, like apparel, accessories and perfumes, and 46.8% on so-called *luxury experience* such as visiting an auction house, staying at a hotel and taking an exclusive trip. A 38.3% share belongs to other luxury items, where the category of food, wines and spirits, for example, contributes to 5.4% (76 billion Euros in market volume) (Mueller-Stewens, 2013). The average annual growth rate of the market for personal luxury goods, luxury cars and experiential luxury between 2010 and 2012 was approximately 13% (Bellaiche et al., 2014). This trend of increasing activities in luxury markets is also apparent in Germany, which is one of the most significant countries of origins for luxury brands in the international context (Meurer, 2012).

In general, the literature considers luxury as a multi-faceted construct (Wiedmann, Hennigs, & Siebels, 2007). The original Latin term *luxus* stands for extravagance, indulgence, affluence and amplitude (Glare, 1992). Furthermore, the "dream value", which signifies the strong desire to consume a specific good, is revealed as decisive characteristic of luxury (Albrecht et al., 2013; Dubois & Paternault, 1995). Even though, the definition of those goods that are luxury for an individual person is highly subjective. For example, luxury can be differentiated between material and immaterial luxury values (Pflanz, 2004). In general, the description of the construct luxury is often based on consumption motives. They are not only influenced by culture, but they are also closely linked with a change in value systems over time (Wong & Ahuvia, 1998). Motives for luxury consumption can generally be subdivided into three categories (Table 1).

These are externally-related, internally-related and finally, motives like price and quality, which are hybrids in this context. Motives for the first category gain importance only in the environment of social structures. They assume the comparison with fellow men and primarily describe the intention to signal social status and milieu affiliation (Han, Nunes, & Dreze, 2010; Wiedmann, Hennigs, & Siebels, 2009). This is closely linked to the concept of *Conspicuous Consumption* described by Veblen (1899). Prestige, status, tradition and self-presentation are further externally-oriented motives that are discussed in the literature (Ascheberg, Meurer, & Oesterling, 2012; Vigneron & Johnson, 2004).

Internally-related motives for luxury consumption result from the desire for indulgence and self-realization. Here, the consumption of luxury goods is the means by which people accomplish hedonic aims and where they live out and refine their personality as well as individuality (Meurer, 2012). The value of a luxury good is generated from an internal-personal sensation or emotional gain. It is therefore uncoupled from the embeddedness of consumers within their social environment (Atwal & Williams, 2009). Tsai (2005) empirically finds that the personal value of luxury consumption is positively correlated with the repurchase intention for a luxury brand.

The third category of hybrid drivers has different functions with regards to the fulfillments of externally-related or internally-related consumption motives, depending on the individual consumer. People associate usability, quality and uniqueness, materialism as well as specific price policies with the value of luxury products (Fassnacht, Kluge, & Mohr, 2013; Hornig, Fischer, & Schollmeyer, 2013; Wiedmann, Hennigs, & Siebels, 2009). They can serve as instruments to not only signal status and prestige but to additionally satisfy personality-based (hedonistic) desires (Vigneron & Johnson, 1999; Yeoman & McMahon-Beattie, 2006).

Table 1. Three categories of motives for luxury consumption

Externally-related motives	Internally-related motives	Hybrid motives
Conspicuous consumption/Signaling	Hedonism and indulgence	Quality
Social stratification	Experience	Usability
Status and prestige	Self-realization	Uniqueness
Tradition	Self-fulfillment	Materialism
Self-portrayal	Individuality	Price
	Inspiration	
	Authenticity	

A change in the motives of Western luxury consumption has been discussed recently. Researchers and practitioners observe that traditional consumption patterns like prestige, social stratification, and conspicuousness, as described by Veblen (1899), take a backseat to modern motives like hedonism, self-expression, and sustainability (Meurer & Manniger, 2012; Pruene, 2012, Yeoman, 2011). Consuming expensive brands and services or accumulating items of property does not primarily serve the externally-related self-image. Instead, according to theory, consumers increasingly satisfy their internal longing by means of consuming luxury goods. The demand for hedonistic actions and individuality becomes a superficial motive for a delightful, luxurious lifestyle in an emancipated, affluent society (Ascheberg, Meurer, & Oesterling, 2012; Atwal & Williams, 2008; Vigneron & Johnson, 2004; Yeoman & McMahon-Beattie, 2012; Yeoman, 2011). As material luxury is widely accessible across social layers in industrial nations, intrinsic desires in contrast to social desires could possibly gain importance (Yeoman & McMahon-Beattie, 2006).

Developments in food consumption in Germany

Similarly to the findings for luxury consumption, there is an observable shift in food consumption motives for some German consumer segments. Product characteristics like quality, sustainability and authenticity, as well as hedonistic values like food enjoyment have gained importance (Institut für Demoskopie Allensbach, 2014), while a low price is decreasingly relevant for purchase decisions (Nestle Deutschland AG, 2012; Nitzko & Spiller, 2014; p. 2006). At the same time, studies reveal that the general significance of nutrition, common meals with family and friends, cooking and enjoying eating in good restaurants is segmentally high (Brunso, Fjord, & Grunert, 2002, Lueth, Spiller, & Enneking, 2004). Brunso, Grunert, and Bredahl (1996) identify "the Adventurous Food Consumers" (AFC), who represent one of five cross-cultural Food-Related Lifestyle segments that are built upon the five dimensions ways of shopping, cooking methods, quality aspects, consumption situations and purchasing motives. The AFC like to cook, are concerned about food quality and associate eating with self-fulfillment. This consumer type, 24% of the German population in 1996, is generally likely to buy gourmet food. (Brunso, Grunert, & Bredahl, 1996; Wycherly, McCarthy, & Cowan, 2008). Lueth, Spiller, and Enneking (2004) further identify two nutrition-conscious segments within Germany: first, people who like to cook and are mainly concerned about health and regionalism of food, and second, those who like to eat in good restaurants, are health- and fitness- oriented, and prefer organic, regional or Fair Trade food. Correspondingly, Kirig and Ruetzler (2007) describe recent food trends in Germany as shifting consumption motives from the former mentality of avariciousness toward health consciousness, quality, hedonistic experience and luxury, and, in accordance to the paragraph above, they postulate that general pleasure markets will be influenced by new luxury values like moral enjoyment, authenticity and health. The authors define food styles as mirrors for trends in society and as indicators for motivational changes in other markets. An explanation for this may be the general embedment of food styles into individual value systems. Brunso, Scholderer, and Grunert (2004) find that the ten motivational domains proposed by Schwartz (1992), covering inter alia hedonism, stimulation and self-direction, are significantly related to the way people consume food.

Overall, the substantially growing demand recognized for organic, regional and Fair Trade food products can be taken as a sign for the establishment of less price-sensitive but more quality-, indulgence-, health- and sustainability-oriented food styles (Baker et al., 2004; Honkanen, Verplanken, & Olsen, 2006; Institut für Demoskopie Allensbach, 2014; Magnusson et al., 2003).

Those identified developments shed new light on premium food products. Major intersections between motives for luxury-consumption and food-consumption imply that food can be associated with luxury and vice versa. Indulgence and quality seem to be in the spotlight of significant consumer segments, presenting a challenge for food marketing. The potential of food as a "new" luxury market, which was established from a trend of trading up goods with an individual emotional content, has already been recognized (Silverstein & Fiske, 2003). Since the composition of a luxury food value has not been empirically investigated to the knowledge of the authors, this study will analyze the effects and significance of the perceived dimensions of luxury food.

Literature review and construct definition

Considerations of luxury food

Approaches to answer the question what does luxury food mean can be found from an archaeological, historical point of view in theoretical literature. Van der Veen (2003) first characterizes luxury food as offering a refinement of a food that is widely desired and second, by means of distinction. She bases the latter on either quantity or quality, symbolizing either success and prestige or exclusivity and distance. Van der Veen (2003) further concludes that there are no specific items of food universally considered to be as a luxury; rather it depends on a particular place, time and society. In developing countries, for example, food quantity is used as a luxury in order to gain prestige, while in complex societies, the emphasis is on food quality for creating exclusivity. Berry (1994) inspires the approach by defining luxury in general as objects of desire, as contradiction to the norm and as instrumental needs. According to him luxury means a refinement or specific quality of a need, e.g. fresh bread instead of bread for satisfying hunger. Needs and desires are not stable over time, they only can be defined in relation to each other and therefore depend on time, stage of development in a society and social layer. Hayden (2003) also addresses the dependence on time, stating "In short, our eating habits today largely are the result of, and reflect, the luxury foods of the past" (p. 459).

Van der Veen (2003) additionally reveals a social component of luxury food as it serves as an instrument for establishing and tightening social relations. Therefore, it is referred to as the meaning of food in feasts. Following Dietler (1990, 1996), there are four categories of feasts where food plays different roles in order to either symbolize exclusiveness or distance, social power or to build up social bonds.

Other contributions towards luxury food are studies on wine. Beverland (2004) state that wine is a suitable example where research on luxury can be done because of three reasons. First, auction prices have increased significantly; second, the evaluation of quality and style are individual and highly variable; and third, premium wine brands, which are defined by means of the product characteristic price, are not marketed by mass marketing. There, three indirect definitions of luxury are found, applied in the context of food: price, quality and marketing. These characteristics defining luxury food may be a reminder on how classical delicacies and premium food is marketed. Following on these results, Beverland (2005) uses wine as basis and investigates another luxury dimension which is authenticity. This is here highlighted as a core component of brand management that becomes increasingly important. Luxury wines are associated with authenticity because they represent the oldest existing brands and are marketed by techniques distanced from industrial products and commerce. Luxury wines are defined in this paper by price and quality. The author investigates the definition of authenticity and concludes that it is "projected via a sincere story that involves the avowal of commitments to traditions, passion for craft and production excellence, and the public disavowal of the role of modern industrial attributes and commercial motivation" (p. 1025). Quality, origin and the decoupling from mass production are found as strong definitional attributes of authenticity. Accordingly, Beverland (2006) highlights heritage and pedigree, stylistic consistency, quality commitments, relationship to place, method of production, and downplaying commercial motives as the six attributes of authenticity, by means of research on wineries and wine consumers. Sidali and Hemmerling (2014) show that authentic food products are associated with more traditional and nostalgic production methods, in contrast to production methods that result from industrialized and intensified agriculture. Furthermore, they find that consumers perceive authenticity as an attribute of small-scale, artisan-produced foods. The image of the producer as an underprivileged actor in niche food markets, in contrast to global food brands, is thereby important for the perceived definition of authentic food. Summarily, origin and the rejection of conventional mass production methods are found again to define authenticity.

Kemp (1998) performs three empirical studies answering the question of whether some food categories are perceived as luxury. Respondents were asked to rate, among other items, wine, bars of chocolate, lunch with friends, fruit juice, fish, milk, bread, a favorite pie in the fridge and champagne on a necessity-luxury scale (with 1=complete necessity and 9=complete luxury). Berry's (1994) differentiation of needs and desires builds the basis of the scale. The results reveal that respondents most frequently associate champagne with luxury (rated 8 on average), followed by wine, bars of chocolate and lunch with friends (rated 7 on average), while fruit juice, fish and favorite pie in the fridge are third-ranked (rated 5 on average), milk is fourth-ranked (rated 3 on average) and bread is least associated with luxury (rated 2 on average). On the one hand, these results confirm the assumption that wine has a high perceived luxury value. On the other hand, the items "lunch with friends" and "favorite pie in the fridge" calls into consideration a new aspect of luxury, which is based on the consumer's personality, on self-directed consumption motives. In conclusion, different aspects of perceived luxury food are found in these studies. The latter food items mirror individual luxury dimensions while champagne, wine and chocolate can be assumed as representatives of quality, price and uniqueness.

Construct definition

With reference to the developments described above, which implicate a definition of luxury food based on a catalogue of modern and classical consumption motives, the concept of this study is based on a motivational, multidimensional approach. Wiedmann, Hennigs, and Siebels (2007, 2009) inspire the structural model for this study, which is extended by the dimension of sustainability and authenticity. This is chosen to represent new luxury attributes in the context of food. In accordance with Jarvis, MacKenzie, and Podsakoff (2003), the mode of measurement is defined to be formative. The causality is from the product attributes to the luxury dimensions, which means that the attributes of a food product determine whether consumers associate a luxury value with it (Figure 1).

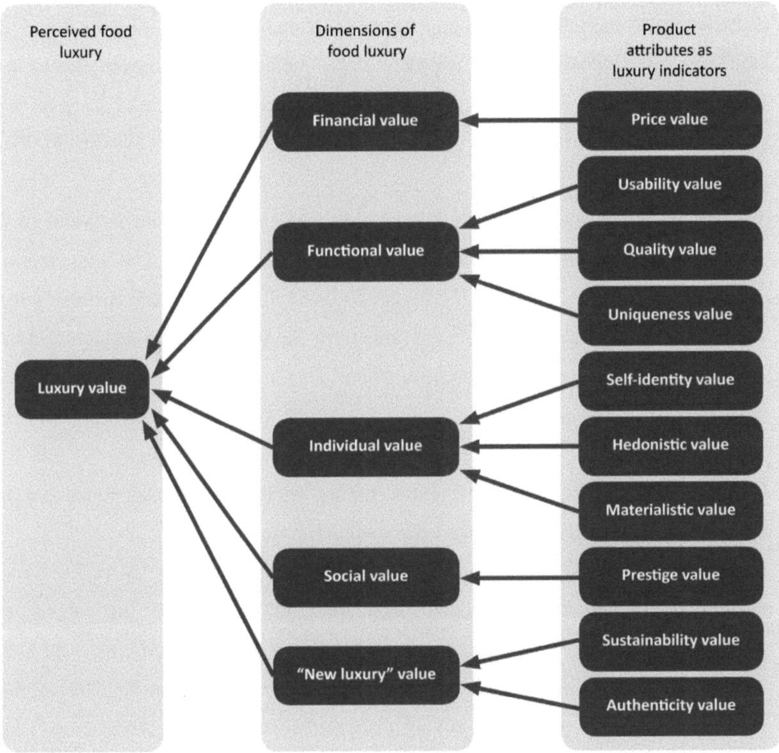

Figure 1. Basic model for investigations[21]

The financial, functional, individual and social values are associated with general luxury, which is evident in literature. High prices are closely connected with perceived quality and exclusivity, and they ultimately act as a moderator variable for the determination of perceived luxury (Erickson & Johansson, 1985; Hornig, Fischer, & Schollmeyer, 2013; Fassnacht, Kluge, & Mohr, 2013; Serraf, 1991; Wiedmann, Hennigs, & Siebels, 2007, 2009; Wheatley & Chiu, 1977). Usability, quality, and uniqueness are revealed to be characteristics of luxury products presenting functionality (Kisabaka, 2001; Vigneron & Johnson, 2004; Wiedmann,

---

[21] Inspired by Wiedmann, Hennigs, and Siebels (2007, 2009).

Hennigs, & Siebels, 2007, 2009; Yeoman & McMahon-Beattie, 2006). So are self-identity, hedonism and materialism accounting for the individual dimension (Dubois & Laurent, 1994; Hudders & Pandelaere, 2012; Puntoni, 2001; Vigneron & Johnson, 2004; Wiedmann, Hennigs, & Siebels, 2007, 2009). The social value can be expressed in terms of prestige (Mason, 1993; Veblen, 1899; Vigneron & Johnson, 1999, 2004; Wiedmann, Hennigs, & Siebels, 2007, 2009).

With regard to the question which attributes are perceived to add luxury value to foods, the dimensions defined for general luxury seem to be applicable. The literature on luxury mentioned above gives evidence for the existence of all four value dimensions: price, quality, self-identity and social meaning. At the same time, the described new emphasis on quality and indulgence in significant German consumer segments reveals the perceived importance of functional and individual values of food. According to these findings, it is hypothesized:

H1: The perceived dimensions of general luxury, which are financial, functional, individual and social, significantly constitute a luxury value for foods.

Recently, consumers have demanded sustainability and authenticity attributes in their food products (Baker et al., 2004; Honkanen, Verplanken, & Olsen, 2006; Institut für Demoskopie Allensbach, 2014; Magnusson et al., 2003; Wenzel, Kirig, & Rauch, 2007). Concurrently, both dimensions are generally actively discussed in conjunction with luxury. Based on a general rise of sustainability concerns in consumer values (Padilla Bravo et al., 2013; Vermeir & Verbeke, 2006), literature shows the challenge for luxury marketing to address concepts like environment friendliness, animal welfare and social sustainability (Bilharz & Belz, 2008; Blevis et al., 2007; Joy et al., 2012; Kapferer, 2010). Authenticity is a crucial instrument in order to create a unique brand identity (Gilmore & Pine, 2007; Sidali & Hemmerling, 2014). Especially in the context of luxury, authenticity is a core element for brands, symbolizing uniqueness, exclusiveness and quality (Beverland, 2005, 2006; Schallehn, 2012). Meurer & Manninger (2012) even postulate it as one of the "Mega-Trends" within the international luxury branch. According to these findings, a second hypothesis is formulated:

H2: Sustainability and authenticity are additional significant dimensions of luxury foods, reflecting new motives in food and luxury consumption.

Methodology

The database was generated with the help of an online questionnaire filled out by 936 German respondents in 2014.[22] The first three items with the highest loadings of each of the factors revealed by Wiedmann, Hennigs, and Siebels (2009)[23, 24] were adapted to the food context and translated into German. For the value of self-directed pleasure, the item "I never buy a luxury food product inconsistent with my own taste." was added. In order to address price, sustainability and authenticity, three further items each were added (Table 3). Respondents evaluated all items on a five-point Likert scale, ranging from full agreement (+2) to no agreement at all (-2). The sample is representative for the German population with regard to the distribution of gender, age, state, net household income per month and highest school leaving certification.

In order to reveal the relevant luxury dimensions for the case of luxury foods, the authors conducted an explorative factor analysis (principal component analysis with the varimax rotation method) of all evaluated items as a first step (using IBM SPSS Statistics 22). Secondly, the authors calculated a hierarchical partial-least-squares path model[25] by using SmartPLS software (Ringle, Wende, & Becker, 2014), taking the revealed factors from the upstream analysis as first-order latent variables. As a second-order latent variable, the luxury value of food items was determined. Using the results of a factor analysis for inspiring a hierarchical PLS model is also done by Wetzels, Odekerken-Schroeder, and van Oppen (2009), who present an application example for PLS path modeling in marketing research. PLS Path Modeling, originally designed by Wold (1974, 1982, 1985) and beneath others extended by Lohmöller (1989), is generally repeatedly applied in marketing sciences and consumer behavior research (Fornell & Robinson, 1983; Hartmann et al., 2013; Henseler, Ringle, & Sinkovics, 2009; Reinartz, Krafft, & Hoyer, 2004). This instrument estimates latent variable scores by means of manifest indicators. It is able to handle complex models with numerous manifest and latent variables and to overcome small sample size problems. The

---

[22] Nine hundred thirty-six respondents remained after data cleansing. The original sample was 1.050 respondents.

[23] With limiting the online questionnaire to the first three items with the highest loadings each, we adapted to time-constraints.

[24] Wiedmann, Hennigs, and Siebels (2009) adopted existing and tested measures (e.g. Dubois & Laurent, 1994; Richins & Dawson,1992; Tsai, 2005) and added items that correspond to results from explanatory interviews.

[25] Path-weighting scheme, 300 maximum iterations, stop criterion value 7, initial outer weights are set to +1, mean replacement to handle missing values.

underlying assumptions for the distribution of data are less stringent. Moreover, reflective as well as formative modes can be estimated and combined in hierarchical model designs. The combination of those characteristics makes it a suitable instrument to analyze causal relationships in empirical consumer studies (Henseler, Ringle, & Sinkovics, 2009).

The variance-based PLS technique is an iterative estimation technique, applying alternating least squares algorithms in order to analyze high dimensional data in a low-structure environment (Henseler, Ringle, & Sinkovics, 2009). It proceeds in three stages. Stage one is subdivided into three steps: First, it estimates outer proxies of latent variable scores as linear functions of their indicators. Second, inner weights, that indicate the intensity of connections between one latent variable to the other latent variables, are estimated. Third, inner proxies of latent variable scores are estimated as linear functions of the outer proxies of their adjacent latent variables, using the aforedetermined inner weights. Last, the outer weights, which are expressed by the covariances between inner proxies of latent variables and its indicators (in reflective model mode), or as the regression weights that result from the ordinary least squares regression of the inner proxi of each latent variable on its indicators (in formative model mode) are determined. In stage two, outer weights/loadings and path coefficients are estimated. Step three is the estimation of location parameters (Henseler, Ringle, & Sinkovics, 2009).

Inner and outer models are differentiated in the PLS technique whereas the first considers relationships among unobserved (or latent) variables and the latter those between latent and their associated observed (manifest) variables. Latent variable scores and all unknown relationships are estimated (Henseler, Ringle, & Sinkovics, 2009).

The modeling technique applied in this study is the repeated-indicators approach where the same manifest variables are used for the first-order latent variables as well as for the second-order latent variable (Lohmüller, 1989; Wold, 1982). To avoid biased results, it is necessary to take the same amount of indicators for every first-order latent variable. Under this condition and for a large number of indicators, the use of the repeated-indicators technique is recommended (Becker, Klein, & Wetzels, 2012; Wilson & Henseler, 2007). Ringle, Sarstedt, & Straub (2012) show that the repeated-indicator approach is often used for hierarchical models in marketing sciences.

Therefore, three items are included for every particular factor in this study, chosen in a way that every aspect from the factors in the principal component analysis is considered. The conduction of a bootstrap test[26] and blindfold test[27] shows the model´s fit and its predictive capability respectively (Hair et al., 2013).

Results

Sample description

Table 2 shows those socio-demographic data that were used as a base for representativeness in comparison to respective distributions within Germany. The data used was gender, age, region, net household income and highest school leaving qualification of the main earner among household members. Additionally, information about the family status, the occupational group and the highest degree of professional education of participants were collected.

Table 2. Comparison between socio-demographic data in the sample and the German population

Socio-demographic data	Sample[1]	German population[2]
Gender	Female: 51.8% Male: 48.2%	Female: 51.0% Male: 49.0%
Age	18–29 years: 16.5% 30–39 years: 15.2% 40–49 years: 21.0% 50–59 years: 22.0% 60–69 years: 18.3% 70–75 years: 7%	18–29 years: 18.7% 30–39 years: 15.9%% 40–49 years: 21.3%% 50–59 years: 20% 60–69 years: 14.7% 70–75 years: 9.4%

---

[26]Bootstrapping with 500 subsamples (basic bootstrapping, no sign changes, two-tailed test, bias-corrected and accelerated bootstrap).

[27] Omission distance: 7.

Region[3]	North: 18.5%	North: 18.2%
	East: 17.5%	East: 17.7%
	West: 35.5%	West: 35.5%
	South: 28.5%	South: 28.6%
Net household income per month	Less than 900€: 8.2%	
	900–1300€: 11.2%	
	1301–1500€: 5.7%	Less than 900€: 8.7%
	1501–2000€: 14%	900–1300€: 11.5%
	2001–2600€: 13.8%	1301–1500€: 5.8%
	2601–3600€: 12.8%	1501–2000€: 14.7%
	3601–5000€: 10.9%	2001–2600€: 14.4%
	5001–7600€: 6.3%	2601–3600€: 17.3%
	7601–9000€: 3.1%	3601–5000€: 14.6%
	9001–12600€: 1.5%	5001–18000 Euro: 13.1%
	12601–18000€: 0.9%	
	More than 18000€: 0.7%	
	No comment: 10.9%	

Highest school leaving qualification of the main earner among household members	Without school leaving qualification: 0.7% Primary school/lower secondary school: 22.1% Qualified degree of lower secondary school: 13.8% Secondary school: 25.4% Polytechnic high school: 6.1% Technical school: 4.0% General or specialized higher education (Abitur): 25.3% Others: 1.7% No comment: 0.9%	Lower Education (primary school, lower secondary school, qualified degree of lower secondary school): 35.6% Middle Education (secondary school, polytechnic high school): 29.1% Higher Education (technical school, general or specialized higher education): 27.3%

[1]N=936.

[2]rounded values; source: Statistisches Bundesamt (2013).

[3]North: Schleswig-Holstein, Hamburg, Mecklenburg-Vorpommern, Lower Saxony, Bremen; East: Berlin, Brandenburg, Saxony-Anhalt, Saxony, Thuringia; West: North Rhine-Westphalia, Hesse, Rhineland-Palatinate, Saarland; South: Bavaria, Baden-Wuerttemberg.

Explorative factor analysis of luxury food dimensions

The factor analysis for luxury motives revealed seven factors that combine the different luxury dimension from the basic model (Figure 1). Seven items were deleted because their loadings were less than 0.5. This rule is stricter than proposed in the literature (Fabrigar et al., 1999; Ford, MacCallum, & Tait, 1986). The Kayser-Meyer-Olkin criterium was 0.889, which is higher than the stated minimum value of 0.6 (Robinson, Shaver &, Wrightsman, 1991). The explained overall variance is 66.086%. Three factors show a Cronbach's Alpha value higher than 0.85 and three one higher than 0.65. One is 0.626. All values are located in the desirable range or are close to it (Bland & Altman, 1997; Nunnally, 1978). The factors are structured as shown in Table 3.

Table 3. Explorative factor analysis[1]

Factor 1: Prestige and Hedonism (Cronbach's Alpha: 0.902, explained share of overall variance: 16.006%)	Factor loading	μ ; σ
I like to know what food products and food brands make good impressions on others.[2]	0.818	-1.11; 0.997
I usually keep up with food style changes by watching what others buy.[2]	0.780	-0.93; 1.061
Before purchasing a food product it is important to know what food brands or products make good impressions on others.	0.777	-1.10; 1.109
Purchasing luxury food provides deeper meaning in my life.[2]	0.726	-1.02; 1.049
Self-realization is an important motivator for my luxury food consumption.	0.724	-0.79; 1.117
When I consume luxury food, cultural development is an important motivator.	0.718	-0.76; 1.022
When in a bad mood, I may consume luxury food for alleviating the emotional burden.	0.606	-0.72; 1.123
Factor 2: Self-identity and Quality (Cronbach's Alpha: 0.791, explained share of overall variance: 11.000%)	**Factor loading**	μ ; σ
I'm inclined to evaluate the substantive attributes and performance of luxury food products myself rather than listen to others' opinions.[2]	0.772	1.18; 0.903
A luxury food product preferred by many people but that does not meet my quality standards will never enter into my purchase consideration.[2]	0.738	1.19; 0.957
Luxury food I buy must match my lifestyle and me.[2]	0.710	0.91; 1.045
I never buy a luxury food product inconsistent with my own taste.	0.702	1.27; 0.982
I buy luxury food for satisfying my personal needs without any attempt to make an impression on other people.	0.668	0.85; 1.227
It is my own desire when I buy a luxury food product not meeting the expectations of others.	0.565	1.16; 0.946
Factor 3: Sustainability and Authenticity (Cronbach's Alpha: 0.857, explained share of overall variance: 9.763%)	**Factor loading**	μ ; σ
Luxury food products should be produced environmentally friendly.[2]	0.820	0.68; 1.096
I only buy a luxury food product if it is produced under fair conditions for humans and animals.[2]	0.813	0.27; 1.163
It is important to me that luxury food products I buy are organic.	0.802	0.03; 1.178
I like to know where the luxury food products I buy are produced.[2]	0.781	0.56; 1.137

Factor 4: Materialism (Cronbach's Alpha: 0.891, explained share of overall variance: 9.045%)	Factor loading	μ ; σ
It sometimes bothers me quite a bit that I can't afford to buy all the luxury food products I'd like.[2]	0.859	-0.58; 1.270
I'd be happier if I could afford to buy more luxury food products.[2]	0.842	-0.68; 1.188
My life would be better if I bought certain luxury food products I don't consume.[2]	0.789	-0.86; 1.111
Factor 5: Usability (Cronbach's Alpha: 0.759; explained share of overall variance: 7.879%)	**Factor loading**	μ ; σ
In my opinion, luxury foods are really useless.[2,3]	0.849	-0.02; 1.117
In my opinion, luxury foods are swanky.[2,3]	0.824	0.34; 1.133
In my opinion, luxury foods are pleasant.[2]	0.591	-0.10; 1.115
Factor 6: Uniqueness (Cronbach's Alpha: 0.626; explained share of overall variance: 6.322%)	**Factor loading**	μ ; σ
True luxury food cannot be mass-produced.[2]	0.793	0.81; 1.001
A luxury food product cannot be sold in supermarkets.[2]	0.751	-0.03; 1.116
Few people consume true luxury food.[2]	0.678	0.74; 1.020
Factor 7: Price (Cronbach's Alpha: 0.651; explained share of overall variance: 6.071%)	**Factor loading**	μ ; σ
I rather buy a certain food product when its price is comparatively high.[2]	0.805	-0.52; 1.034
When buying food, the price is not decisively.[2,3]	-0.747	0.38; 1.050
I associate high prices for food products with particularly high quality.[2]	0.573	-0.08; 1.087

[1] Items evaluated with five-point Likert scale (from -2=full disagreement to 2=full agreement); Principal component analysis with varimax rotation and Kaiser normalization; 7 iterations, Kayser-Meyer-Olkin criterion: 0.889; Bartlett significance: 0.000; cumulated explained share of overall significance: 66.086%.

[2] Included in the Partial Least Squares Path Modeling.

[3] Reverse coded items.

Partial least squares structural equation analysis

According to Jarvis, MacKenzie, and Podsakoff (2003), the direction of causality in a PLS path model and the correlation or non-correlation of measures are criteria in the choice of the mode. Reflective and formative constructs are further differentiated by the interchangeability of indicators. Reflective constructs imply that removing one indicator does not change the nature of the underlying construct. In formative models, the latent variable is instead defined by its measures and therefore, they are necessary parts in the construct (Diamantopoulos &

Winklhofer, 2001). By means of these three criteria, the mode of the model was chosen. First, the causality goes from the factors to the luxury value in the inner model, and from the factors to the items in the outer model. Second, the indicators for the first-order latent variables are expected to be (highly) correlated. The principal component analysis has shown that they can be summarized to seven first-order factors, meaning that the respective indicators for each factor are loading on the same shared luxury dimension. The items related to the same factor in the outer model are per se significantly correlated on the level 0.01 (Wolff & Bacher, 2010). The outer model thus should be measured in the reflective mode. For the inner model instead, we do not have reasons to assume that the first-order latent variables, the luxury product attributes (compare Figure. 1), are correlated. Factors that are revealed in a principal component analysis are orthogonal (Comon, 1994).[28] The second-order construct should thus be modeled by the formative mode (Jarvis, MacKenzie, & Podsakoff, 2003).

Last, the removal of one of the factors in the inner model means omitting one of the luxury dimensions that defines the variable luxury value. Contrarily, removing one of the items loading on the same factor would not question the existence of the respective luxury dimension. Taking these findings into consideration, a reflective-formative model mode, as described by Becker, Klein, and Wetzels (2012) was chosen (Figure 2). Correspondingly, in the reflective part of the model, the manifest variables are a linear function of their associated latent variables, and vice versa in the formative part (Henseler, Ringle, & Sinkovics, 2009). Other applications of reflective-formative constructs in consumer behavior research are shown by Fornell & Robinson (1983) and Hartmann et al. (2013).

---

[28] The orthogonality of the factors in the principal component analysis has been empirically checked, using the Pearson correlation coefficient. It yielded that the factors are empirically uncorrelated.

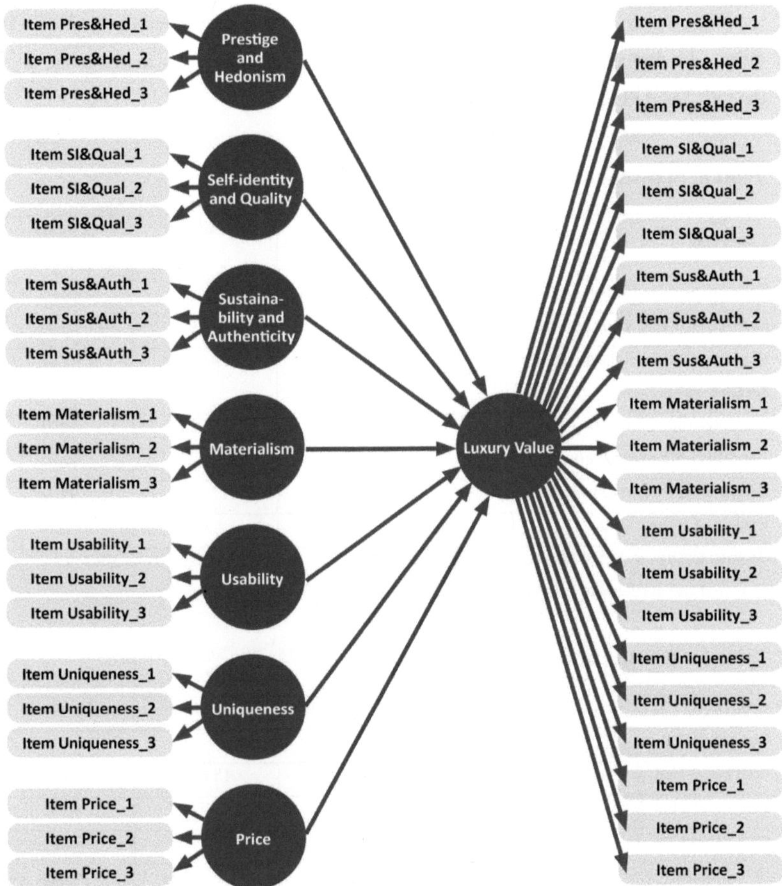

Figure 2. Path model

The predictive relevance of the structural model is given for every latent variable. A blind-folding analysis (omission distance of 7) shows that all values for Stone-Geisser's Q^2 as a measure for cross-validated redundancy are greater than zero (Fornell & Bookstein, 1982; Geisser, 1974; Stone, 1974). For the second-order construct (Luxury Value), the Q^2 is 0.699. The regression analysis shows that Self-identity and Quality, Uniqueness, Usability and Materialism have the strongest effect on the perceived luxury value of a food (0.187; 0.186; 0.186; 0.185). The weakest coefficient is the estimate for Price (0.017). Prestige and Hedonism is in the middle, but with a path coefficient closer to the first ones (0.170) (Table 4). The outer loadings, except the items for Price in the second-order construct satisfy the Chin (1998) condition of being greater than 0.6. All path coefficients and the outer loadings are significant at the 0.01 level (Tables 4 and 5).

Table 4. Statistics of the inner model

Contructs	Path coefficients	T statistics[1]	P value[1]
Prestige and Hedonism → Luxury Value	0.170	15.730	<0.01
Self-identity and Quality → Luxury Value	0.187	15.792	<0.01
Sustainability and Authenticity → Luxury Value	0.168	13.539	<0.01
Materialism → Luxury Value	0.185	18.028	<0.01
Usability → Luxury Value	0.186	16.903	<0.01
Uniqueness → Luxury Value	0.186	17.625	<0.01
Price → Luxury Value	0.017	4.502	<0.01

[1]Bootstrapping with 500 subsamples (basic bootstrapping, no sign changes, two-tailed test, bias-corrected and accelerated bootstrap).

Table 5. Statistics of the outer model

Contructs	Outer loading	T statistics[1]	P value[1]
Luxury Value →			
Item Pres&Hed_1	0.850	15.634	<0.01
Item Pres&Hed_2	0.888	18.521	<0.01
Item Pres&Hed_3	0.887	18.376	<0.01
Item SI&Qual_1	0.947	20.167	<0.01
Item SI&Qual_2	0.944	19.043	<0.01
Item SI&Qual_3	0.945	19.281	<0.01
Item Sus&Auth_1	0.849	15.570	<0.01
Item Sus&Auth_2	0.849	15.637	<0.01
Item Sus&Auth_3	0.849	15.667	<0.01
Item Materialism_1	0.935	27.969	<0.01
Item Materialism_2	0.934	28.035	<0.01
Item Materialism_3	0.931	27.603	<0.01

Item Usability_1	0.942	19.640	<0.01
Item Usability_2	0.940	19.128	<0.01
Item Usability_3	0.945	20.459	<0.01
Item Uniqueness_1	0.938	19.525	<0.01
Item Uniqueness_2	0.942	20.421	<0.01
Item Uniqueness_3	0.941	20.371	<0.01
Item Price_1	0.098	3.513	<0.01
Item Price_2	0.114	4.885	<0.01
Item Price_3	0.124	3.719	<0.01
Prestige and Hedonism →			
Item Pres&Hed_1	0.961	33,205	<0.01
Item Pres&Hed_2	0.990	177.181	<0.01
Item Pres&Hed_3	0.990	177.062	<0.01
Self-identity and Quality →			
Item SI&Qual_1	0.996	247.004	<0.01
Item SI&Qual_2	0.995	25.123	<0.01
Item SI&Qual_3	0.995	28.981	<0.01
Sustainability and Authenticity →			
Item Sus&Auth_1	0.998	1,203.316	<0.01
Item Sus&Auth_2	0.998	1,273.093	<0.01
Item Sus&Auth_3	0.998	1,344.362	<0.01
Materialism →			
Item Materialism_1	0.998	875.727	<0.01
Item Materialism_2	0.998	994.697	<0.01
Item Materialism_3	0.997	685.312	<0.01
Usability →			
Item Usability_1	0.996	113.628	<0.01
Item Usability_2	0.995	87.351	<0.01
Item Usability_3	0.993	156.985	<0.01
Uniqueness →			
Item Uniqueness_1	0.994	69.638	<0.01
Item Uniqueness_2	0.995	101.938	<0.01
Item Uniqueness_3	0.994	83.331	<0.01
Price →			
Item Price_1	0.782	19.765	<0.01
Item Price_2	0.740	16.204	<0.01
Item Price_3	0.774	22.035	<0.01

[1]Bootstrapping with 500 subsamples (basic bootstrapping, no sign changes, two-tailed test, bias-corrected and accelerated bootstrap).

The outer loadings are greater than 0.9 for first-order constructs, except for Price where they range from 0.740 and 0.782. For the second-order construct, all outer loadings are above 0.85, except for Price where they range from 0.098 and 0.124. All outer loadings are again significant on the level 0.01.

The Cronbach's Alpha values for the first-order latent variables, except for Price, are above 0.7 and thus again in the desirable range (Nunnally, 1978). In accordance with Hair, Ringle, and Sarstedt (2011), the reflective outer model is evaluated by four criteria. The average variance extracted is above 0.5 for all latent constructs, which means that the convergent validity of the model is given. The composite reliability measure shows high internal consistency reliability since all values are greater than 0.7 (Hair, Ringle, & Sarstedt, 2011) (Table 6).

Table 6. Evaluation criteria of the outer model

Latent variables	Cronbach's Alpha (CA)[1]	Average Variance Extracted (AVE)[1]	Composite Reliability (CR)[1]
Luxury Value	0.973	0.720	0.980
Prestige and Hedonism	0.980	0.961	0.987
Self-identity and Quality	0.995	0.991	0.997
Sustainability and Authenticity	0.998	0.995	0.998
Materialism	0.998	0.995	0.998
Usability	0.995	0.989	0.996
Uniqueness	0.994	0.988	0.996
Price	0.649	0.586	0.810

[1] All values are significant on the level 0.001 in bootstrapping analysis.

Indicator loadings in a reflecctive model mode that are higher than 0.7 imply their reliability (Hair, Ringle, & Sarstedt, 2011). This is given for all indicator loadings, except for Price in the second-order construct. Last, the discriminant validity of the construct is measured by means of cross-loadings and the Fornell-Larcker criterion (in accordance with Henseler, Ringle, & Sinkovics, 2009). Checking the cross-loadings should reveal that every indicator has a higher correlation with its respective latent variable than with another latent variable in the model (Henseler, Ringle, & Sinkovics, 2009). This is given for this construct. All correlation of the indicators with their respective latent variable are close to one. The indicators for Price have the lowest correlations with their latent variable, but are still higher than the respective cross-loadings. They range from 0.782 to 0.740. The Fornell-Larcker criterion shows discriminant validity for this model since the square root of the average variance extracted is higher for each latent construct than the squared correlation with any other latent construct (Fornell & Larcker, 1981; Georges & Eggert, 2003) (Table 7).

Table 7. Discriminant validity: Fornell-Larcker Criterion[1]

	1	2	3	4	5	6	7
1. Prestige and Hedonism	**0.980**						
2. Materialism	0.840	**0.998**					
3. Price	0.149	0.129	**0.766**				
4. Self-identity and Quality	0.726	0.853	0.096	**0.996**			
5. Sustainability and Authenticity	0.955	0.722	0.121	0.677	**0.998**		
6. Uniqueness	0.722	0.849	0.097	0.987	0.670	**0.994**	
7. Usability	0.724	0.855	0.136	0.983	0.669	0.978	**0.995**

[1]Bold numbers show the square root of the AVE, numbers below the diagonal represent squared construct correlations.

Table 7 further shows that on the one hand, some of the latent first-order constructs are highly correlated, e.g. Uniqueness and Self-identity/Quality (squared correlation of 0.987) or Prestige/Hedonism and Self-identity/Quality and Usability (squared correlation of 0.983). On the other hand, the squared correlations of Price and other constructs are rather low (the highest with Usability which is 0.136). This supports for all first-order constructs except Price the idea that they measure the same underlying latent variable, namely Luxury Value. Price as a characteristic of food products seems to play a distinctive role. It is not only little correlated with other dimensions, it also shows a small effect on Luxury Value in the regression (path coefficient of 0.017).

Discussion

In this study, the question to what particular extent certain luxury dimensions constitute a perceived luxury food value is answered. Empirical investigations are based on data that was generated by means of a representative questionnaire of 936 German consumers conducted in the summer of 2014. As an underlying concept, a multidimensional framework inspired by Wiedmann, Hennigs, and Siebels (2007) is chosen. Accordingly, luxury is defined by using financial, functional, individual and social values. For the analysis, two different instruments are applied. First, the principal component analysis determines the relevant luxury food dimensions. Seven factors are found: Prestige and Hedonism, Self-identity and Quality, Sustainability and Authenticity, Materialism, Usability, Uniqueness and Price. In a second step, using the factors as first-order latent variables (including only the first three items of each of the factors), a hierarchical partial least squares path model is estimated in order to reveal the dimensions' effect on a luxury food value. A reflective-formative model mode was chosen due to decision rules specified by Jarvis, MacKenzie, and Podsakoff (2003).

It was found that all dimensions significantly contribute to a perceived luxury food value. Self-identity and Quality has the highest path coefficient (0.187), closely followed by Uniqueness and Usability (0.186) and Materialism (0.185), Hedonism/Prestige and Sustainability/Authenticity range next (0.170; 0.168) while for Price, the weakest coefficient, is estimated (0.017). Consequently, the latter has to be differentiated from the others: The comparison of squared correlations shows that Price is the only dimension that correlates only weakly with the other latent constructs. Regardless of possible methodological difficulties (see the limitations in the next chapter), this result implies that price is less associated with a luxury value than the other first-order constructs.

The revealed definitional dimensions of a luxury food value correspond to the dimensions that are found for a general luxury value. The latter can be substantially subsumed under a financial luxury value (price), functional luxury values (uniqueness, quality and usability), individual values (hedonism, self-identity, materialism) as well as social values (prestige and conspicuousness) (Dubois, Laurent, & Czellar, 2001; Tsai, 2005; Vigneron & Johnson, 2004; Wiedmann, Hennigs, & Siebels, 2009). Therefore, luxury food can be clearly identified as a luxury market. Similar and significant path coefficients for the six constructs other than Price confirm for luxury foods what is implicated by literature on general luxury marketing: luxury value is perceived to be multi-faceted (Vigneron & Johnson, 2004; Wiedmann, Hennigs, & Siebels, 2007, 2009). In the environment of luxury food, functional and individual facets are

190

the most significant. This is not only congruent with the postulated change in motives for luxury consumption, it also mirrors the recently found (new) emphasis on quality and self-identity (e.g. by means of self-made meals) in the German food sector (Institut für Demoskopie Allensbach, 2014, Kirig & Ruetzler, 2007; Lueth et al., 2004; Nestle Deutschland AG, 2012).

Moreover, the significance of new luxury values in the form of sustainability and authenticity can be confirmed. This result also corresponds to the postulated shift of motives for general luxury consumption toward individual, internally-oriented values, functionality as well as sustainability and authenticity concerns (Ascheberg, Meurer, & Oesterling, 2012; Atwal & Williams, 2008; Meurer & Manniger, 2012; Pruene, 2013; Yeoman, 2011; Yeoman & McMahon-Beattie, 2010).

In summary, both hypotheses of this study cannot be rejected. Even though, the dimension of Price contributes very weakly to a luxury food value and the dimension of Sustainability and Authenticity is not among the most relevant dimensions, all path coefficient are significant.[29] An explanation for the lower effect of Sustainability and Authenticity could be that luxury and sustainability are still perceived to be opposites by some consumer segments. According to Kapferer (2010), this attitude is based on the segmental found assumption that luxury serves as an instrument to create inequality within the society. If the individual perception of luxury is mainly based on traditional, externally-oriented dimensions, reluctance toward prestige and conspicuous consumption will mean that sustainable and authentic attributes of luxury food will be denied. Padilla Bravo et al. (2013) find another possible reason. They consider that marketing strategies subsumed under "exclusiveness" could threaten organic food marketing. Consumers might associate exclusive gourmet and specialties with conventional production methods. This would delete the link between sustainable, authentic foods and luxury as far as gourmet and specialties are defined as luxury food.

Interestingly, the path coefficient for the dimensions except price are similar. This can be explained by the fact that the luxury value is virtually the mean of the 18 items used to measure the dimensions Prestige and Hedonism,..., Uniqueness (compare Figure 2, left), whereas the items related to price only have low loadings (Table 5). As all dimensions are measured using the same number (3) of indicators, all dimensions, except price, contribute

---

[29] It should mentioned here that even small effects can be significant with a sufficient large sample size (Becker, Klein, & Wetzels, 2012). Therefore, it cannot be excluded that the large sample size is responsible for the significant path coefficient of Price.

equally to the luxury value. This is typical for the repeated-indicators approach with the number of indicators for each dimension being equal (Becker, Klein, & Wetzels, 2012). In the study by Zhang, Li, and Sun (2006), where the highest difference in numbers of indicators is only one, similar results can be observed.

Another methodological weakness of the repeated-indicator approach is the following: The first-order and higher-order latent variables are modelled as linear combinations of the same indicators. Therefore, the entire variance of the latter is explained by the first, what causes an R^2-value of one. Some authors refer to this problem (Becker, Klein, & Wetzles, 2012; Ringle, Sarstedt, & Straub, 2012; Wetzels, Odekerken-Schroeder, & van Oppen, 2009).

Limitations and next research steps

In this paper, two instruments for empirical statistics, the explorative factor analysis and the PLS path modeling, are combined in order to investigate the perceived dimensions of luxury in the context of one particular product category, namely foods. It contributes to the literature and to practical marketing since the definition of luxury is still under-researched, even though an upcoming significance of international luxury markets is recognized (Hagtvedt & Patrick, 2009). As far as the authors know, this study is one of the first empirical contribution that regards the definition of luxury foods.

With regard to its limitations, the applied methodology accounts for similar path coefficients and an R^2-value of one (as explained in the discussion). For the conceptual model of this study, it is thus recommendable to test the hybrid approach, where each indicator is used only once (Marsh, Wen, & Hau, 2006; Wilson & Henseler, 2007), or the two-stage approach, that is based on subsequently using latent factor scores for the estimation of higher-order components (Wilson & Henseler, 2007). Agarwal & Karahanna (2000) show an application of the two-stage technique. Fornell and Robinson (1983) as well as Hartmann et al. (2013) provide applications of reflective-formative models with hybrid indicators in consumer behavior research. Applying the hybrid approach on luxury food would require the generation of additional indicators that can be used to model the luxury value. This is desirable for future research.

Additionally, the role of price should be further analyzed. The effect of the Price on Luxury value is rather small in this study (Table 4). At the same time, the quality of this latent variable as part of the model is low. Values for CA, AVE and CR as well as the squared correlations with all other latent constructs are comparatively low (Tables 6 and 7). Furthermore, the outer loadings in the second-order construct are lower than 0.7 (the benchmark for the indicators'reliability in reflective models according to Ringle & Sarstedt, 2011) (Table 5), and the cross-loading criterion show the weakest results for Price among all factors. The choice of other items included in the factor analysis, e.g. with stronger links to luxury food (compare the applied items in Table 3) could improve the explanatory power of the financial dimension. However, Wiedmann, Hennigs, and Siebels (2009) already reveal the specificity of price in defining luxury value. They find that price acts as a moderator variable signaling the consumer that a product contains other luxury values, e.g. quality. They thus do not find price as a separate luxury dimension.

Furthermore, it was useful to modify the model for other luxury markets, taking new consumption patterns into consideration. Here, luxury foods are investigated and accordingly, the sustainability and authenticity were added to the original dimensions. Other relevant new luxury dimensions could be immaterial values like education, having time for family and friends, strong social relationships or the "dream value" (Dubois & Paternault, 1995; Weber & Dubois, 1997).

Last, it might be interesting to investigate the dimensions of luxury (food) empirically with an international database. Both luxury and food consumptions patterns differ among cultures and nations (Gracia & Albisu, 2001; Wong & Ahuvia, 1998). Against this background, it should be checked if luxury food is differently defined according to the culture under investigation.

References

Agarwal, R., & Karahanna, E. (2000). Time flies when you're having fun: Cognitive absorption and beliefs about information technology usage. *MIS quarterly, 24*(4), 665-694.

Albrecht, C.-M., Backhaus, C., Gurzki, H., & Woisetschlaeger, D. M. (2013). Value Creation for Luxury Brands through Brand Extensions: An Investigation of Forward and Reciprocal Effects. *Marketing ZFP – Journal of Research and Management, 35*(2), 91-103.

Ascheberg, C., Meurer, J., & Oesterling, A. (2012). The Luxury Universe–Angebots-und Kundensegmentierung globaler Luxusmärkte als Basis für erfolgreiche Positionierungsstrategien. In C. Burmann, V. König, & J. Meurer (Hrsg.), *Identitätsbasierte Luxusmarkenführung* (S. 85-101). Springer Fachmedien Wiesbaden.

Atwal, G., & Williams, A. (2009). Luxury brand marketing–the experience is everything!. *Journal of Brand Management, 16*(5), 338-346.

Baker, S., Thompson, K. E., Engelken, J., & Huntley, K. (2004). Mapping the values driving organic food choice: Germany vs the UK. *European Journal of Marketing, 38*(8), 995-1012.

Becker, J. M., Klein, K., & Wetzels, M. (2012). Hierarchical latent variable models in PLS-SEM: guidelines for using reflective-formative type models. *Long Range Planning, 45*(5), 359-394.

Bellaiche, J., Kluz, M., Mei-Pochtler, A., & Wiederin, E. (2012). Luxe Redux: raising the bar for selling of luxuries. *Boston Consulting Group, Boston.*

Berry, C. J. (1994). *The idea of luxury: A conceptual and historical investigation* (Vol. 30). Cambridge University Press.

Beverland, M. B. (2004). Uncovering "theories-in-use": building luxury wine brands. *European Journal of Marketing, 38*(3/4), 446-466.

Beverland, M. B. (2005). Crafting brand authenticity: the case of luxury wines. *Journal of Management Studies, 42*(5), 1003-1029.

Beverland, M. B. (2006). The 'real thing': Branding authenticity in the luxury wine trade. *Journal of Business Research, 59*(2), 251-258.

Bilharz, M., & Belz, F. M. (2008). Öko als Luxus-Trend: Rosige Zeiten für die Vermarktung „grüner " Produkte?. *Marketing Review St. Gallen, 25*(4), 6-10.

Bland, J. M., & Altman, D. G. (1997). Statistics notes: Cronbach's alpha. *Bmj, 314*(7080), 572.

Blevis, E., Makice, K., Odom, W., Roedl, D., Beck, C., Blevis, S., & Ashok, A. (2007, August). Luxury & new luxury, quality & equality. In *Proceedings of the 2007 conference on Designing pleasurable products and interfaces* (pp. 296-311). ACM.

Brunsø, K., Grunert, K. G., & Bredahl, L. (1996). *An analysis of national and cross-national consumer segments using the food-related lifestyle instrument in Denmark, France, Germany and Great Britain.* Aarhus, Denmark: MAPP.

Brunsø, K., Scholderer, J., & Grunert, K. G. (2004). Testing relationships between values and food-related lifestyle: results from two European countries. *Appetite, 43*(2), 195-205.

Chin, W. W. (1998) The partial least squares approach to structural equation modelling. In G. A. Marcoulides (Ed.), *Modern methods for business research* (pp. 295–336). Mahwah, NJ: Lawrence Erlbaum.

Comon, P. (1994). Independent component analysis, a new concept?. *Signal processing, 36*(3), 287-314.

D´arpizio, C., & Levato, F. (2014). Lens on worldwide luxury consumer: Relevant segments, behaviors and consumption patterns, nationalities and generations compared. Published by Bain and Company, Boston.

Diamantopoulos, A., & Winklhofer, H. M. (2001). Index construction with formative indicators: an alternative to scale development. *Journal of Marketing Research, 38*(2), 269-277.

Dietler, M. (1990). Driven by drink: the role of drinking in the political economy and the case of Early Iron Age France. *Journal of Anthropological Archaeology, 9*(4), 352-406.

Dietler, M. (1996). Feasts and commensal politics in the political economy: Food, power, and status in prehistoric Europe. In P. Wiessner & W. Schiefenhoevel (Eds.), *Food and the status quest: An interdisciplinary perspective* (pp. 87-125). Providence, RI: Berghahn.

Dubois, B., & Laurent, G. (1994). Attitudes toward the concept of luxury: An exploratory analysis. *Asia-Pacific Advances in Consumer Research, 1*(2), 273-278.

Dubois, B., & Paternault, C. (1995). Observations: Understanding the world of international luxury brands: The" dream formula". *Journal of Advertising Research, 35*(4), 69-76.

Dubois, B., Laurent, G., & Czellar, S. (2001). *Consumer rapport to luxury: Analyzing complex and ambivalent attitudes* (No. 736). HEC Paris.

Erickson, G. M., & Johansson, J. K. (1985). The role of price in multi-attribute product evaluations. *Journal of Consumer Research, 12*(2), 195-199.

Fabrigar, L. R., Wegener, D. T., MacCallum, R. C., & Strahan, E. J. (1999). Evaluating the use of exploratory factor analysis in psychological research. *Psychological methods, 4*(3), 272.

Fassnacht, M., Kluge, P. N., & Mohr, H. (2013). Pricing Luxury Brands: Specificities, Conceptualization and Performance Impact. *Marketing ZFP – Journal of Research and Management, 35*(2), 104-117.

Ford, J. K., MacCallum, R. C., & Tait, M. (1986). The application of exploratory factor analysis in applied psychology: A critical review and analysis. *Personnel Psychology, 39*(2), 291-314.

Fornell, C., & Bookstein, F. L. (1982). Two structural equation models: LISREL and PLS applied to consumer exit-voice theory. *Journal of Marketing Research, 19*(4), 440-452.

Fornell, C., & Larcker, D. F. (1981). Evaluating structural equation models with unobservable variables and measurement error. *Journal of Marketing Research, 18*(1), 39-50.

Fornell, C., & Robinson, W. T. (1983). Industrial organization and consumer satisfaction/dissatisfaction. *Journal of Consumer Research, 9*(4), 403-412.

Geisser, S. (1974). A predictive approach to the random effect model. *Biometrika, 61*(1), 101-107.

Georges, L., & Eggert, A. (2003). Key account managers' role within the value creation process of collaborative relationships. *Journal of Business to Business Marketing, 10*(4), 1-22.

Gilmore, J. H., & Pine, B. J. (2007). *Authenticity: What consumers really want* (Vol. 1). Boston, MA: Harvard Business School Press.

Glare, P. G. (1982). *Oxford latin dictionary*. Clarendon Press. Oxford University Press.

Gracia, A., & Albisu, L. M. (2001). Food consumption in the European Union: Main determinants and country effects. *Agribusiness, 17*(4), 469-488.

Hagtvedt, H., & Patrick, V. M. (2009). The broad embrace of luxury: Hedonic potential as a driver of brand extendibility. *Journal of Consumer Psychology, 19*(4), 608-618.

Hair Jr, J. F., Hult, G. T. M., Ringle, C., & Sarstedt, M. (2013). *A primer on partial least squares structural equation modeling (PLS-SEM)*. SAGE Publications, Incorporated.

Hair, J. F., Ringle, C. M., & Sarstedt, M. (2011). PLS-SEM: Indeed a silver bullet. *The Journal of Marketing Theory and Practice, 19*(2), 139-152.

Han, Y. J., Nunes, J. C., & Drèze, X. (2010). Signaling status with luxury goods: the role of brand prominence. *Journal of Marketing, 74*(4), 15-30.

Hartmann, L., Kerssenfischer, F., Fritsch, T., & Nguyen, T. (2013). User Acceptance of Customer Self-Service Portals. *Journal of Economics, Business and Management, 1*(2), 150-155.

Hayden, B. (2003). Were luxury foods the first domesticates? Ethnoarchaeological perspectives from Southeast Asia. *World Archaeology, 34*(3), 458-469.

Henseler, J., Ringle, C. M., & Sinkovics, R. R. (2009). The use of partial least squares path modeling in international marketing. In R. R. Sinkovics, P. N. Ghauri (Eds.), *New Challenges to International Marketing (Advances in International Marketing, Volume 20)* (pp. 277-319). Emerald Group Publishing Limited.

Honkanen, P., Verplanken, B., & Olsen, S. O. (2006). Ethical values and motives driving organic food choice. *Journal of Consumer Behaviour, 5*(5), 420-430.

Hornig, T., Fischer, M., & Schollmeyer, T. (2013). The Role of Culture for Pricing Luxury Fashion Brand. *Marketing ZFP – Journal of Research and Management, 35*(2), 118-130.

Hudders, L., & Pandelaere, M. (2012). The silver lining of materialism: the impact of luxury consumption on subjective well-being. *Journal of Happiness Studies, 13*(3), 411-437.

Institut für Demoskopie Allensbach (2014).Vertrauen und Skepsis: Was leitet die Deutschen beim Lebensmitteleinkauf?. *SGS Germany GmbH. Hamburg.*

Jarvis, C. B., MacKenzie, S. B., & Podsakoff, P. M. (2003). A critical review of construct indicators and measurement model misspecification in marketing and consumer research. *Journal of Consumer Research, 30*(2), 199-218.

Joy, A., Sherry, J. F., Venkatesh, A., Wang, J., & Chan, R. (2012). Fast fashion, sustainability, and the ethical appeal of luxury brands. *Fashion Theory: The Journal of Dress, Body & Culture, 16*(3), 273-296.

Kapferer, J. N. (2010). All that glitters is not green: The challenge of sustainable luxury. *European Business Review*, 40-45.

Kemp, S. (1998). Perceiving luxury and necessity. *Journal of Economic Psychology, 19*(5), 591-606.

Kirig, A., & Ruetzler, M. H. (2007). *Food-Styles. Die wichtigsten Thesen, Trends und Typologien für die Genuss-Märkte.* Zukunftsinstitut-Studie. Kelkheim.

Kisabaka, L. (2001). *Marketing für Luxusprodukte* (Vol. 32). Dissertation, Fördergesellschaft Produkt-Marketing, Cologne.

Lohmoeller, J. B. (1989). *Latent variable path modeling with partial least squares* (p. 130). Heidelberg: Physica-Verlag.

Lueth, M., Spiller, A., & Enneking, U. (2004). *Analyse des Kaufverhaltens von Selten- und Gelegenheitskäufern und ihrer Bestimmungsgründe für/gegen den Kauf von Öko-Produkten.* Projektabschlussbericht für das BMVEL im Rahmen des Bundesprogramms ökologischer Landbau, Goettingen.

Magnusson, M. K., Arvola, A., Hursti, U. K. K., Åberg, L., & Sjoedén, P. O. (2003). Choice of organic foods is related to perceived consequences for human health and to environmentally friendly behaviour. *Appetite, 40*(2), 109-117.

Marsh, H. W., Wen, Z., Hau, K-T. (2006). Structural Equation Models of Latent Interaction and Quadratic Effects. In G. R. Hancock, & R. O. Mueller (Eds.), *Structural Equation Modeling: A Second Course* (pp. 225-268). Information Age Publishing, Greenwich.

Mason, R. (1993). Cross-cultural influences on the demand for status goods. *European Advances in Consumer Research, 1*, 46-51.

Meurer, J. (2012). Ebony or Ivory–wie glänzend ist die Zukunft des Luxus in Deutschland? Kritische Reflexionen zum Luxusmarkenmanagement. In C. Burmann, V. König, & J. Meurer (Hrsg.), *Identitätsbasierte Luxusmarkenführung* (S. 321-336). Springer Fachmedien Wiesbaden.

Meurer, J., & Manninger, K. (2012). Quo vadis globale Luxusmarkenführung–Status, Trends und Top-Themen für die CMO-Agenda. In C. Burmann, V. König, & J. Meurer (Hrsg.), *Identitätsbasierte Luxusmarkenführung* (S. 13-31). Springer Fachmedien Wiesbaden.

Mueller-Stewens, G. (2013). Das Geschäft mit Luxusgütern. Universität St. Gallen.

Nestlé Deutschland AG (2012) (Hrsg.). *Nestlé Studie 2012, Das is(s)t Qualität*, Auszüge aus der Nestlé Studie 2012.

Nitzko, S., & Spiller, A. (2014). Zielgruppenansätze in der Lebensmittelvermarktung, In M. Halfmann (Hrsg.), *Zielgruppen im Konsumentenmarketing* (S. 315-332). Springer Fachmedien Wiesbaden.

Nunnally, J. C., & Bernstein, I. H. (1978). Psychometric theory. New York: McGraw-Hill.

Padilla Bravo, C., Cordts, A., Schulze, B., & Spiller, A. (2013). Assessing determinants of organic food consumption using data from the German National Nutrition Survey II. *Food Quality and Preference, 28*(1), 60-70.

Page, C. (2006). *Innovation in Gourmet and Specialty Food and Drinks: Market Evolution and NPD in Super-premium and Healthy Products.* Business Insights Limited.

Pflanz, C. (2003). Faszination Luxus. In C. Pflanz (Hrsg.), *Faszination* (S. 90-97). Gabler Verlag.

Pruene, G. (2013). *Luxus und Nachhaltigkeit.* Springer Fachmedien Wiesbaden.

Puntoni, S. (2001). Self-identity and purchase intention: An extension of the theory of planned behavior. *European Advances in Consumer Research, 5*, 130-134.

Reinartz, W., Krafft, M., & Hoyer, W. D. (2004). The customer relationship management process: its measurement and impact on performance. *Journal of Marketing Research, 41*(3), 293-305.

Richins, M., & Dawson, S. (1992). A consumer values orientation for materialism and its measurement: Scale development and validation. *Journal of Consumer Research, 19*(3), 303–316.

Ringle, C. M., Sarstedt, M., & Straub, D. W. (2012). Editor's comments: a critical look at the use of PLS-SEM in MIS Quarterly. *MIS Quarterly, 36*(1), iii-xiv.

Ringle, C. M., Wende, S., & Becker, J.-M. (2014). *SmartPLS 3.* Hamburg: http://www.smartpls.com.

Robinson, J. P., Shaver, P. R., & Wrightsman, L. S. (1991). Criteria for scale selection and evaluation. *Measures of personality and social psychological attitudes, 1*(3), 1-16.

Schallehn, M. (2012). Identitätsbasierte Führung von Luxusmarken unter besonderer Berücksichtigung der Marken-Authentizität am Beispiel von Bugatti und Maybach. In C. Burmann, V. König, & J. Meurer (Hrsg.), *Identitätsbasierte Luxusmarkenführung* (S. 53-67). Springer Fachmedien Wiesbaden.

Schwartz, S. H. (1992). Universals in the content and structure of values: theoretical advances and empirical tests in 20 countries. In M. P. Zanna (Eds.), *Advances in experimental social psychology* (vol. 25, pp. 1–65). San Diego, CA: Academic Press.

Serraf, G. (1991). Le produit de luxe: somptuaire ou ostentatoire?. *Revue française du marketing*, (132), 7-16.

Sidali, K. L., & Hemmerling, S. (2014). Developing an authenticity model of traditional food specialties: Does the self-concept of consumers matter?. *British Food Journal, 116*(11), 1692-1709.

Silverstein, M., & Fiske, N. (2003). *Trading Up*. New York: Portfolio.

Statistisches Bundesamt (2013). Zahlen und Fakten. URL: https://www.destatis.de/DE/ZahlenFakten/GesellschaftStaat/StaatGesellschaft.html;jsessio nid=C96AEA9A9BCDEF2DAA17DF50700EB5A6.cae1. Last accessed: 31 January 2015.

Stone, M. (1974). Cross-validatory choice and assessment of statistical predictions. *Journal of the Royal Statistical Society. Series B (Methodological), 36*(2), 111-147.

Tsai, S. P. (2005). Impact of personal orientation on luxury-brand purchase value. *International Journal of Market Research, 47*(4), 429-454.

Van der Veen, M. (2003). When is food a luxury?. *World Archaeology, 34*(3), 405-427.

Veblen, T. (1899). *The theory of the leisure class*. Oxford University Press.

Vermeir, I., & Verbeke, W. (2006). Sustainable food consumption: Exploring the consumer "attitude–behavioral intention" gap. *Journal of Agricultural and Environmental Ethics, 19*(2), 169-194.

Vigneron, F., & Johnson, L. W. (1999). A review and a conceptual framework of prestige-seeking consumer behavior. *Academy of Marketing Science Review, 1*(1), 1-15.

Vigneron, F., & Johnson, L. W. (2004). Measuring perceptions of brand luxury. *The Journal of Brand Management, 11*(6), 484-506.

Weber, D., & Dubois, B. (1997). The edge of the dream: Managing brand equity in the European luxury market. In L. R. Kahle, & L. Chiagouris, (Eds.), *Values, lifestyles, and psychographics* (pp. 263-282). Lawrence Erlbaum Associates, Inc.

Wenzel, E., Kirig, A., & Rauch, C. (2007). *Zielgruppe LOHAS: Wie der grüne Lifestyle die Märkte erobert.* Zukunftsinstitut-Studie, Kelkheim.

Wetzels, M., Odekerken-Schröder, G., & Van Oppen, C. (2009). Using PLS path modeling for assessing hierarchical construct models: guidelines and empirical illustration. *MIS quarterly, 33*(1), 177-195.

Wheatley, J. J., & Chiu, J. S. (1977). The effects of price, store image, and product and respondent characteristics on perceptions of quality. *Journal of Marketing Research, 14*(2), 181-186.

Wiedmann, K. P., Hennigs, N., & Siebels, A. (2007). Measuring consumers' luxury value perception: a cross-cultural framework. *Academy of Marketing Science Review, 7*(7), 333-361.

Wiedmann, K. P., Hennigs, N., & Siebels, A. (2009). Value□based segmentation of luxury consumption behavior. *Psychology & Marketing, 26*(7), 625-651.

Wilson, B., & Henseler, J. (2007). Modeling reflective higher-order constructs using three approaches with PLS path modeling: a Monte Carlo comparison. ANZMAC.

Wold, H. O. (1974). Causal flows with latent variables: partings of the ways in the light of NIPALS modeling. *European Economic Review, 5*(1), 67-86.

Wold, H. O. (1982). Soft modeling: The basic design and some extensions. In K. G. Joereskog & H. O. Wold (Eds.), *Systems under indirect observations*, Part II (pp. 1–54). Amsterdam: North-Holland.

Wold, H. O. (1985). Partial least squares. In S. Kotz, & N. L. Johnson (Eds), *Encyclopedia of statistical sciences.* (Vol. 6, pp. 581–591). New York, NY: Wiley.

Wolff, H. G., & Bacher, J. (2010). Hauptkomponentenanalyse und explorative Faktorenanalyse. In C. Wolf, & H. Best, *Handbuch der sozialwissenschaftlichen Datenanalyse* (pp. 333-365). VS Verlag für Sozialwissenschaften.

Wong, N. Y., & Ahuvia, A. C. (1998). Personal taste and family face: Luxury consumption in Confucian and Western societies. *Psychology and Marketing, 15*(5), 423-441.

Wycherley, A., McCarthy, M., & Cowan, C. (2008). Speciality food orientation of food related lifestyle (FRL) segments in Great Britain. *Food quality and preference, 19*(5), 498-510.

Yeoman, I. (2011). The changing behaviours of luxury consumption. *Journal of Revenue & Pricing Management, 10*(1), 47-50.

Yeoman, I., & McMahon-Beattie, U. (2006). Luxury markets and premium pricing. *Journal of Revenue and Pricing Management, 4*(4), 319-328.

Yeoman, I., & McMahon-Beattie, U. (2010). The changing meaning of luxury. In I. Yeoman and U. McMahon-Beattie (Eds.), *Revenue Management: A Practical Pricing Perspective* (chapter 6, pp. 62-85), Basingstoke, UK: Palgrave MacMillan.

Zhang, P., Li, N., & Sun, H. (2006, January). Affective quality and cognitive absorption: Extending technology acceptance research. In *System Sciences, 2006. HICSS'06. Proceedings of the 39th Annual Hawaii International Conference on* (Vol. 8, pp. 207a-207a). IEEE.

Chapter III: Excursus: Marketing Instruments in a Branch that is Associated with Luxury

III.1   Weiterentwicklung des Rankings im Reitsport – ein Experiment

Autoren: Christian Schulz-Wiemann, Laura Hartmann, Achim Spiller, Jan Gertheiss

Georg-August-Universität Göttingen

Dieser Beitrag wurde in ähnlicher Form in der Zeitschrift *Sportwissenschaft, 44*(2), 99–115, veröffentlicht.

Zusammenfassung

Die kritische Hinterfragung der bestehenden Rankingsysteme im deutschen und internationalen Reitsport kann Schwachstellen hinsichtlich ihrer Aussage zum Leistungsniveau von Sportpferden aufdecken. Ziel der vorliegenden Studie ist es, aus der derzeit verwendeten kumulativen Erfolgsdarstellung eine leistungsgerechtere und dennoch realisierbare Bewertungsmethodik zu entwickeln. Die Modellierung geschieht zum einen auf Basis eines Durchschnittsverfahrens, zum anderen unter Berücksichtigung von Zusatz- und Minuspunkten, wobei die Ergebnisse der neuen Rankings anhand einer Stichprobe von fünfzig Top-Springpferden mit den bestehenden Rankings verglichen werden. Daraus ergeben sich deutliche Veränderungen in der Rangfolge. Durch die Umstellung kann die Validität in Bezug auf die Leistungsbeurteilung der Pferde maßgeblich verbessert werden. Überdies erfahren die verschiedenen Anspruchsgruppen eine stärkere Berücksichtigung ihrer Interessen. Daher empfiehlt es sich, das bestehende Rankingsystem entsprechend weiterzuentwickeln und eine differenziertere Punktevergabe eingehender zu analysieren.

Schlüsselwörter

Sportpferde – Springreiten – Leistungsbeurteilung – Ranking – Rangliste

Abstract

A critical analysis of the existing ranking systems of German and international horse sports could identify weak points with regard to their implicit evaluations of the horses' performances. This study is aimed at developing a feasible and more performance-related valuation method and comparing it with the existing cumulative ranking systems of the German and the international equestrian umbrella associations. On the one hand, the modeling is based on a method using average points per competition; on the other hand, by considering additional points and drawbacks. We investigated a sample of fifty top-ranked jumping horses and monitored the differences toward the existing ranking systems. The modification proposed results in significant implications on the rankings. The validity in relation to the horses' performance evaluation could be improved. As a consequence, the interests of different stakeholder groups are better served than before. It is recommended to further develop the existing ranking system accordingly and to analyze a more differentiated placing of points.

Keywords

Sport horses, Show Jumping, Performance Evaluation, Ranking, Ranking List

Einleitung

Rankingsysteme im Sport

Der Vergleich sportlicher Leistungen mit der Zielsetzung der Leistungskontrolle und Transparenz stellt den Ursprung von Rankings dar (Stölting, 2002). Neben der Dokumentation der Leistung entscheiden Rangordnungen heute im Sport über Auf- oder Abstieg, die Teilnahme an bestimmten Turnieren und die Bereitstellung finanzieller Mittel (Frisby, 1986; West, 2006). Darüber hinaus werden sie als Variable in Modellen zur Vorhersage zukünftiger sportlicher Leistungen eingesetzt (Raab & Philippen, 2008). Im Reitsport sollen sie hauptsächlich Schlüsse auf die sportliche Leistungsfähigkeit der Pferde zulassen und damit den entsprechenden Zielgruppen zuverlässige Informationen bereitstellen. Inwieweit die derzeit angewendeten Rankings dieser Anforderung gerecht werden, wird hier geprüft. Das derzeit angewendete Bewertungsverfahren im deutschen Reitsport beschränkt sich auf die Kumulation der Erfolge, d.h. der Top-Platzierten einer abgeschlossenen Saison. Daraus ergibt sich, dass mindestens zwei Drittel der erbrachten Leistungen in Bezug auf alle Pferde gesehen keinerlei Berücksichtigung finden (FN, 2011). Die kritische Hinterfragung des Rankingsystems deckt Möglichkeiten der Optimierung auf. Ein alternatives Modell mit Einbeziehung aller Leistungen und einer differenzierteren Punktevergabe wird am Beispiel von 50 Top-Springpferden untersucht und mit dem aktuellen Rankingsystem verglichen. Ziel ist ein leistungsgerechteres und aktuelleres Ranking der Pferde, welches dem Sportverband und den weiteren Zielgruppen wie Reitern, Züchtern, Besitzern, Käufern und Verkäufern sowie Zuchtverbänden und Medien eine Ergänzung zum bestehenden System bietet oder dieses ablösen könnte.

In Literatur und Praxis wird teilweise zwischen „Rating" und „Ranking" unterschieden, wobei mit „Rating" häufig die eigentliche (subjektive) Bewertung eines Objekts und die Zuweisung eines Ergebniswertes gemeint sind. Die anschließende ordinale Platzierung wird dann als „Ranking" bezeichnet (Stefani, 2011). Kladroba (2005), der in den meisten veröffentlichten „Ratings" und „Rankings" eine Mischform aus beidem sieht, nimmt die Abgrenzung über die Art der Bewertung vor: Führt diese zu einem absoluten Ergebnis (gut/schlecht), handelt es sich um ein Rating. Ist sie dagegen relativer Natur (besser/schlechter), spricht er von Ranking. Entsprechend wird hier die Bewertung als integraler Bestandteil des Rankings betrachtet.

Die Beschreibung der Systematik zu den unterschiedlichen Rankings im Sport kann angelehnt werden an Stefani (2011): *Kumulative Rankingsysteme,* wie sie derzeit im deutschen Reitsport angewendet werden, beinhalten die Summation der Ergebnisse j des Sportlers bzw. Pferdes i in einem festgelegten Zeitraum:

$$x_i = \sum_j f_{ij} (Ergebnisse\ in\ Punkten; ggf.\ Gewichtung; ggf.\ weitere\ Faktoren)$$

Ausgangspunkt ist in der Regel die Platzierung des Sportlers in einem Wettkampf, die in Punkte umgerechnet wird. Bei einigen Sportarten werden zusätzlich der Schwierigkeitsgrad des Wettkampfs durch den Einsatz von Gewichtungsfaktoren gewürdigt oder Faktoren wie anteilige Punkte aus nicht jährlich stattfindenden Meisterschaften oder Gewinnsummen mit einbezogen. Kumulative Rankings setzen den Anreiz, an möglichst vielen Wettkämpfen teilzunehmen. So sollen Eintrittskarten besser verkauft und Einschaltquoten hoch gehalten werden.

Bei den *adjustiven Rankings* vergleicht man entweder das aktuelle Ergebnis mit einem Prognosewert, der aus vergangenen Ergebnissen ermittelt wurde, oder man bildet einen Durchschnitt. Diesem System lassen sich Elo-, Probit- und Durchschnittsmethoden zuordnen (Chen et al., 2002; Elo, 1978;).

Subjektive Rankings basieren auf voneinander unabhängig erstellten Bewertungen mehrerer Experten und einer anschließenden Berechnung des Gesamtergebnisses pro Kontrahent.

Anwendungsfeld Reitsport

Der Reitsport nimmt eine Sonderstellung im Sport ein, da dort ein Lebewesen im Mittelpunkt steht (Löwe, 1988). Verglichen werden können entsprechend Sportler oder Pferde. Der folgende Beitrag bezieht sich auf die Leistungsfähigkeit der Tiere. Das Pferd ist durch viele, zum Teil nicht objektiv messbare Eigenschaften charakterisiert. Seine Leistungsfähigkeit hängt von diversen Faktoren ab, so sind Haltung, Fütterung, Charakter, Trainingsbedingungen, Wetter und Starterfeld auf einem Turnier sowie das Zusammenpassen mit dem Reiter wichtige Einflussgrößen (Arnemann, 2003; Arnold, 2010; Mamerow, 2010). Da eine Berücksichtigung aller inneren und äußeren Faktoren nicht möglich ist, wird im Folgenden von optimalen Bedingungen für die Pferde ausgegangen.

Der nationale Reitsportverband, die *Deutsche Reiterliche Vereinigung* (FN), bietet zwei Verfahren an, um Sportpferde in eine Rangfolge zu bringen. Im ersten Verfahren bekommen die Pferde für das erfolgreiche Abschließen einer Prüfung Ranglistenpunkte (RP). Die erreichbaren RP sind von der FN für die unterschiedlichen Schwierigkeitsklassen aller Disziplinen definiert (FN, 2013c). Von allen Pferden, die in einer Springprüfung starten, wird mindestens das erste Viertel platziert, Punkte gibt es maximal für das erste Drittel (FN, 2013a). Ein genauer Anteil an zu platzierenden Pferden ist nicht vorgeschrieben. Diese vage Definition verursacht Schwankungen in der Anzahl der Platzierten im Verhältnis zu den Startern – je nach Prüfung und Turnier. Im zweiten Verfahren findet eine Kumulation der Gewinnsummen der Pferde statt. Das erste Viertel der Platzierten bekommt im Regelfall eine entgeltliche Entlohnung (FN, 2013a). Die Ranglistenpunkte und die Gewinnsumme werden für die beendete Saison zusammengezählt. Darüber hinaus bildet das System Lebensgewinnsummen ab. Ziel ist in erster Linie, eine Grundlage für die monetäre Beurteilung von Pferden sowie für die Vergabe von Startberechtigungen in bestimmten Prüfungen zu schaffen. Eine weitere wichtige Verwendung finden die Sportergebnisse in der Zuchtwertschätzung (Dohms-Warnecke, 2012). Besonders erfolgreiche Pferde sollen als Vererber in der Zucht eingesetzt werden, um Eigenschaften wie ihre körperliche und mentale Leistungsfähigkeit zu streuen.

Die *Internationale Reiterliche Vereinigung* (FEI), der internationale Dachverband, ermittelt für alle internationalen Reitturniere und Meisterschaften ebenfalls die RP aus den Ergebnissen, die in Abhängigkeit von Gewinnsummen und Regionen in Matrixform gestaffelt werden (FEI, 2012c). Auf Jahresbasis stehen Ranglisten für die einzelnen Regionalgebiete (z.B. Nordeuropa) und ein globales Ranking zur Verfügung (FEI, 2012b). Die Daten der FEI verwendet man ebenfalls für die Zuchtwertschätzung. Sie werden an die *World Breeding Federation for Sport Horses* (WBFSH) weitergeleitet, wo sie in Zucht- und Sportrankings für Pferde einfließen (WBFSH, 2013).

Neben Abweichungen hinsichtlich der Anzahl der im Ranking berücksichtigten Prüfungen und ihrer Wertigkeit können die in Tab. 1 dargestellten, hier relevanten Unterschiede zwischen den Rankingmethoden der FN und FEI identifiziert werden (FEI, 2013; FEI, 2012a; FN, 2013a; FN, 2013c).

Tab 1. Unterschiede in den Rankingmethoden der FN und der FEI

FEI (international)	FN (national)
RP für die besten 16 Teilnehmer einer Prüfung	RP für das beste Viertel bzw. das beste Drittel einer Prüfung (keine einheitliche Regelung)
Abstufung der erhaltenen RP über alle 16 Plätze	Gleiche Anzahl RP für alle Teilnehmer ab dem 6. Platz
Generell weniger RP als bei der FN	Generell mehr RP al bei der FEI
Bereitstellung vollständiger Ergebnislisten von den Prüfungen eines Turniers	Keine Bereitstellung vollständiger Ergebnislisten
Kostenlose Bereitstellung von Rankings für Reiter sowie Pferd-Reiter-Kombinationen (FEI, 2013)	Kostenpflichtige Bereitstellung des Pferderankings (FN, 2013d) und kostenlose Bereitstellung der besten 2.000 Reiter einer jeden Saison (FN, 2013b)
Abgesehen vom Zeitfenster keine Auswahlmöglichkeiten für die Ordnung der Rankings nach weiteren Kriterien	Ordnung der Rankings nach Rasse, Zuchtgebiet und Alter des Pferdes möglich (zeitlich auf das Vorjahr begrenzt)

Problemstellung

Ein Messinstrument muss die Hauptgütekriterien der Objektivität, Reliabilität und Validität erfüllen, um die Aussagefähigkeit der Ergebnisse zu gewährleisten (Hammann & Erichson, 2000; Himme, 2006). Da das Ranking im Reitspringsport anhand von klar definierten Parametern (Punktematrix, Ergebnislisten) erfolgt, sind Objektivität und Reliabilität per se gegeben. Fraglich ist jedoch das Vorliegen von Validität (Heasman et al., 2008): Wird mit der kumulativen Berücksichtigung der jeweils erreichten RP tatsächlich das gemessen, was auch gemessen werden sollte? Und wird das Nebengütekriterium der Vollständigkeit nicht dadurch massiv verletzt, dass ein Großteil der Ergebnisse der Pferde nicht einbezogen wird?

Im Jahr 2011 wurden auf 3.594 Turnierveranstaltungen in Deutschland 67.750 Prüfungen mit 1.459.828 Starts ausgetragen. Bei diesen ca. 1.500.000 Starts wurden ca. 500.000 Pferde platziert. Rund 1 Mio. Ergebnisse wurden nicht berücksichtigt (FN, 2011). Folglich können gute, aber unplatzierte Ergebnisse nicht erfasst und damit die Sporttauglichkeit von Pferden nur teilweise abgebildet werden. Eine hohe Transparenz und Aussagekraft des Rankings erfordern jedoch eine Gesamtbetrachtung, d. h. zur fairen, sachgerechten und damit validen Beurteilung der erbrachten Leistungen müssen alle Ergebnisse im Betrachtungszeitraum einbezogen werden. Nur so können auch sinnvolle Rückschlüsse auf die künftigen Leistungen eines Pferdes im Sport oder eines Vererbers in der Zucht gezogen werden. Während im Sport noch sehr häufig eine Beschränkung auf positive Ergebnisse erfolgt (Stefani, 2011), ist es in anderen Lebensbereichen üblich, auf Grundlage der Gesamtheit aller Ergebnisse eine Leistungsbeurteilung vorzunehmen. Beispielsweise fließen in das Abschlusszeugnis eines Universitätsabsolventen alle Prüfungsergebnisse ein, und die Stiftung Warentest veröffentlicht neben den guten ebenfalls die als mangelhaft getesteten Produkte. Auch nehmen bereits einige Sportverbände eine Gesamtbetrachtung durch Elo-, Probit- oder Durchschnittsmethoden mindestens auf internationaler Ebene vor.

Das Ranking der FEI lässt durch die konstante Berücksichtigung der ersten 16 Teilnehmer einer Prüfung sowie die differenzierte Punktevergabe im Vergleich zur FN eine transparentere und umfassendere, wenn auch nicht vollständige Beurteilung der Leistungen eines Pferdes zu. Dieser Beitrag entwickelt Modifikationen der bestehenden Rankings im Reitsport und prüft diese auf ihre Eignung zur Optimierung der Validität. Als Besonderheit des Beispiels Reitsport erweist sich dabei die Schwierigkeit, das Leistungsvermögen von Pferden zu messen (d.h. fehlender Goldstandard) sowie das Vorhandensein von Diskrepanzen in den Leistungen von breitensportlich und spitzensportlich eingesetzten Turnierpferden. Zur Einschränkung des Umfangs konzentrieren wir unsere Messungen in dieser Studie auf den Spitzensport. Dennoch lassen sich auf den Ergebnissen basierend auch Überlegungen für den Breitensport anstellen. Hierbei wird angenommen, dass breitensportlich eingesetzte Pferde in weniger schweren Prüfungen starten und häufiger aus dem Parcours ausscheiden.

Möglichkeiten zur objektiven Leistungsmessung in Bezug auf verschiedene andere Sportarten werden in der Literatur diskutiert. Jeremic und Radojicic (2010) führten ein ähnliches Experiment durch, um zwei verschiedene kumulative, im Turnierschach eingesetzte Rankings hinsichtlich ihrer Güte zu evaluieren. Sie identifizieren Schwächen der aktuellsten Modifikation im Schach-Ranking und schlagen die ursprünglich für das Ranking von Ländern nach Entwicklungsstand entwickelte I-Distanzmethode als validere Alternative für die Rangierung von Schach-Teams vor.

Erwägungen für den Fußballsport sind häufig der Verbesserung von Spielprognosen gewidmet und beziehen sich nur indirekt auf die Beurteilung einzelner Sportler. Gerhards, Mutz und Wagner (2012) schlagen vor, Prognosen für Welt- und Europameisterschaften an Marktwerte von Spieler und Mannschaften zu knüpfen. Bei sehr ausgeglichenen Kontrahenten wäre aber auch die Zufallskomponente entscheidend. Letzteres wird zudem bei Heuer und Rubner (2012a) und Heuer, Müller und Rubner (2010) thematisiert. Torschüsse können dargestellt werden als Poisson-verteilte Zufallsprozesse (unter der Prämisse verschiedener Teamstärken).

Vergangene Spielminuten bzw. vorausgegangene Tore beeinflussen die Eintrittswahrscheinlichkeit nicht.

Heuer und Rubner (2009, 2012b) stellen fest, dass Tordifferenzen die Mannschaftsleistung valider abbilden als die Anzahl der erspielten Punkte bzw. Torschüsse und sich damit besser als Grundlage für Prognosen eignen. Des Weiteren können sie zeigen, dass von Konstanz der Mannschaftsfitness innerhalb einer Spielsaison auszugehen ist. Rue und Salvesen (2000) problematisieren die gegenseitige Abhängigkeit der Offensiv-Defensiv-Stärke von konkurrierenden Teams und entwickeln ein Bayesianisches Prognosemodell, das der Abhängigkeit Rechnung trägt und zudem retrospektive Spielanalysen zulässt.

Mukherjee (2012) sowie Radicchi (2011) entwickeln mithilfe von sozialer Netzwerkanalyse Rankings im Profitennis bzw. Cricket, die trotz ihrer Unabhängigkeit von externen Kriterien, wie subjektive Bewertungskomponenten, die Qualität einzelner Siege berücksichtigen. Siege gegen erfolgreichere Konkurrenten werden höher gewichtet als Siege gegen weniger erfolgreiche Konkurrenten. Stochastische Prozesseigenschaften werden auch in Baseball-Spielen identifiziert. Sire und Redner (2009) zeigen, dass sich Siege und Niederlagen, über einen langen Zeitraum betrachtet, wie Zufallsprozesse mit nur geringem Selbstverstärkungsprozess verhalten.

West (2006) liefert einen Überblick über bestehende Rankings im *NCAA*-Basketball und entwickelt ein neues System, mit welchem sich die an *NCAA*-Turnieren teilnehmenden Mannschaften einfacher, flexibler und effektiver rangieren lassen. Dieses System basiert auf logistischer Regression und der Bildung von Erwartungswerten. In Tab. 2 werden die hier zitierten Studien dokumentiert.

Tab. 2. Literaturübersicht

Autoren	Zweck der Leistungsmessung	Erkenntnisgewinn für die Leistungsmessung im Sport	Legitimation der Validität
Heuer, Müller und Rubner (2010)	Prognose von Spielausgängen im Fußball	Torschüsse in einem Spiel sind Zufallsprozesse	Torschüsse in einem Spiel lassen sich als Poisson-Prozess darstellen (bei Berücksichtigung der differierenden Fitness verschiedener Teams)
Gerhards, Mutz und Wagner (2012)	Prognose des Ausgangs von Meisterschaften und Turnieren im Fußball	1. Marktwerte erweisen sich als sinnvoll für die Prognose. 2. Je ausgeglichener die Mannschaften (z.B. in Liga), umso größer der Einfluss von Faktoren wie Glück und Tagesform (Zufall) bzw. Schiedsrichterurteile auf den Spielausgang	1. Zusammenhang zwischen Mannschafts-Marktwerten am Saisonbeginn und erspielten Punkten am Saisonende feststellbar ($r=0{,}73$). 2. Richtige Voraussagen der Marktwertmethode in vergangenen Meisterschaften 3. Welt- und Europameisterschaften besser prognostizierbar als Meisterschaft in Liga
Heuer und Rubner (2012a)	Prognose des weiteren Spielverlaufs aus dem vergangenen Spielverlauf	Torschüsse in einem Spiel sind Poisson-verteilt (bei Berücksichtigung der differierenden Fitness verschiedener Teams)	Messung konstanter Torschusswahrscheinlichkeiten innerhalb eines Spiels
Heuer und Rubner (2012b)	Prognose von Tordifferenzen als Spielausgang im Fußball	Anzahl an Torchancen ist gegenüber der Anzahl geschossener Tore aussagekräftiger	Bessere Prognosequalität von Torchancen
Heuer und Rubner (2009)	Statistische Spezifizierung von Ergebnissen in einer Fußballliga	1. Tordifferenzen eignen sich besser als Instrument zur Messung der Mannschaftsfitness als Anzahl der Punkte 2. Mannschaftsfitness innerhalb einer Saison ist nahezu konstant	1. Bessere Prognose der Ergebnisse einer zweiten Saisonhälfte aus Informationen der ersten Hälfte 2. Unabhängigkeit der Team-Fitness vom Zeitpunkt innerhalb einer Saison kann gezeigt werden
Jeremic und Radojicic	Definition eines „Goldstandards" für Rankings im Turnierschach	I-Distanzmethode eignet sich besser für das Schach-Ranking als die angewende-	Einbezug von 15 Performance-Variablen in I-Distanzmethode im Ver-

(2010)		te, an Fußball angelehnte Methode (3-1-0-Regel)	gleich zu zwei Variablen in 3-1-0-Regel
Mukherjee (2012)	Modifizierung des bestehenden Rankingsystems im Cricket zur Identifikation der besten Teams und Kapitäne	Netzwerkanalyse unter Verwendung des Page-Rank-Algorithmus ist ein effektives Instrument zur Entwicklung des Team- bzw. Teamkapitän-Rankings im Cricket	Page-Rank-Algorithmus berücksichtigt die „Qualität" von Siegen (Höhergewichtung von Siegen gegen stärkere Mannschaften) und ist unabhängig von externen Faktoren
Radicchi (2011)	Entwicklung eines historischen Rankings im Profitennis mithilfe von sozialer Netzwerkanalyse	Entwicklung des „prestige score" als unabhängige Variable im Ranking zur Bestimmung der Qualität von Siegen	Neues Rankingsystem liefert bessere Prognosen als herkömmliches System (basierend auf Anzahl erreichter Punkte pro Spieler pro Saison) und ist unabhängig von externen Faktoren
Rue und Salvesen (2000)	Entwicklung eines Prognosemodells für den Fußball unter Einbeziehung der gegenseitig voneinander abhängigen Mannschaftsstärken	Bereitstellung eines Prognosemodells, das die gegenseitige Abhängigkeit von Offensiv-Defensiv-Teamstärken im Zeitverlauf berücksichtigt und retrospektive Analysen zulässt	1. Vergleiche zwischen Prognosen und Quoten von Buchmachern zeigen gute Übereinstimmung 2. Retrospektive Wahrscheinlichkeiten für finale Rankings sind effektiv berechenbar
Sire und Redner (2009)	Beschreibung der statistischen Eigenschaften von Siegen und Niederlagen im Baseball	Siege und Niederlagen, über einen langen Zeitraum betrachtet, können als Zufallsprozesse mit nur geringen Selbstverstärkungsprozess beschrieben werden	Bradley-Terry-Modell (berücksichtigt nur relative Stärke von konkurrierenden Teams) prognostiziert Siege bzw. „Erfolgs- und Pechphasen" einzelner Teams sehr gut
West (2006)	1. Evaluation bestehender Rankings im NCAA Basketball 2. Entwicklung eines einfacheren und flexibleren Systems	Ranking auf Grundlage der OLRE-Methode ist aussagekräftiger, einfacher und flexibler als zwei bestehende Basketball-Rankings (RPI und Computer Rating nach Jeff Sagarin)	Bessere Prognosen durch OLRE als Bradley-Terry-Simulation auf Grundlage von Daten aus Rating nach Jeff Sagarin

OLRE Ordinal Logistic Regression Modeling and Expectation

Methode

Entwicklung der Hypothesen

Im Folgenden soll anhand von zwei Beispielen überprüft werden, wie sich die Bildung von Punkte-Durchschnitten und die Vergabe von Zusatz- und Minuspunkten auf Basis von unplatzierten Prüfungsergebnissen auf das bestehende Ranking auswirken. Im ersten Schritt wird ein Durchschnitt der RP pro Turnierstart gebildet. Dadurch erfolgt eine implizite Berücksichtigung aller Leistungen eines Pferdes. Im Anschluss wird überprüft, ob die Berücksichtigung von Aufgabe und Ausscheiden das Ranking beeinflusst. Insbesondere das Ausscheiden könnte ein wichtiger Indikator dafür sein, ob ein Pferd den Anforderungen einer Schwierigkeitsklasse gewachsen ist oder ob es etwa charakterliche Mängel (Unwilligkeit) aufweist. Um das „knappe Scheitern" von Pferd und Reiter zu honorieren, werden Zusatzpunkte für diejenigen Pferde vergeben, die zu den besten 50% gehören und nicht platziert wurden. Beides sind Ansätze, um die heutige Gleichbehandlung aller nicht platzierten Teilnehmer weiter zu differenzieren.

Die vereinfachten Beispiele sind realitätsnah gehalten: Die Werte wurden frei gewählt, entsprechen jedoch einer auf langjähriger Erfahrung beruhenden Einschätzung. Für die Punktevergabe wurde die Annahme getroffen, dass in jeder Prüfung dreißig Pferde teilnahmen und jeweils ein Drittel platziert wurde. Das heißt, die ersten zehn Pferde erhielten jeweils RP. Für die Platzierungen in den jeweiligen Kategorien wurden die RP gemäß den Bestimmungen der FN vergeben (FN, 2013c). In Beispiel 1 wird neben der Gesamtpunktzahl der Durchschnitt „RP pro Start" unter Berücksichtigung der Anzahl der Starts gebildet (Tab. 3). Die gleiche Anzahl an Platzierungen und auch die ähnliche Gesamtpunktzahl erwecken den Eindruck, dass die zwei Pferde auf einem ähnlichen Leistungsniveau sind. Es lässt sich mutmaßen, dass Pferd B das leistungsstärkere ist. Unter Berücksichtigung der Startanzahl ändert sich der Eindruck. Pferd A hat für das Erreichen der 98 RP fünf Turnierstarts benötigt, Pferd B hingegen ist auf 15 Turnieren angetreten, um 108 RP zu sammeln. Die „RP pro Start" zeigen nun, dass Pferd A mit 19,6 RP pro Start wesentlich besser einzuschätzen ist als Pferd B mit 7,2 RP pro Start.

Tab. 3. Beispiel 1 – Ermittlung von durchschnittlichen Punkten pro Start

Pferd	A	B
Anzahl der Platzierungen	4	4
Anzahl der Starts	5	15
Ergebnisse in Kategorie L (=leichte Klasse)[1]	2, 4, 8, 12	1, 5, 1, 3, 18, 29, 17, 22, 24, 13
Ergebnisse in Kategorie M (= mittelschwere Klasse)[1]	7	28, 19, 24, 16, 17,
RP Gesamt[2]	98	108
Rang	2	1
RP pro Start	19,6	7,2
Rang neu	1	2

[1] Die Ergebnisse sind ausgedrückt in Form von Platzierungen innerhalb einer Prüfung (2 = 2. Platz usw.)

[2] Die Anzahl der jeweiligen zu vergebenen Punkte pro Platzierung ist angelehnt an das Reglement der FN

In Beispiel 2 werden Pferde, die sich in der ersten Hälfte des Teilnehmerfeldes befinden, aber nicht platziert wurden, mit der Hälfte der Punktzahl des Letztplatzierten bewertet, das Ausscheiden mit dem negativen Wert in dieser Höhe (Tab. 4). Aus Gründen der Übersichtlichkeit verzichten wir hier auf das Bilden des Durchschnitts.

Nach Summation der regulären FN-Punkte schneidet Pferd C mit 132 RP deutlich besser ab als Pferd D mit 72 RP. Da Pferd C aber in drei Springen der Kategorie M ausgeschieden ist, werden hier jeweils 17,5 RP – die Hälfte der RP des Letztplatzierten – von der Gesamtsumme abgezogen. Pferd D hingegen erhält für die Plätze 14 in Kategorie L sowie für 11 und 13 in Kategorie M insgesamt 42,5 Punkte zusätzlich. Der erneute Vergleich zeigt, dass nun Pferd D mit 114,5 RP besser einzuordnen ist als Pferd C mit 79,5 RP.

Tab. 4. Beispiel 2 – Vergabe von Zusatz- und Minuspunkten

Pferd	C	D
Ergebnisse in Kategorie L[1]	3, 1, 2, 1, 10	2, 4, 3, 14, 19, 25
Ergebnisse in Kategorie M[1]	a, a, a[2], 28, 19	13, 11, 16, 20
RP Gesamt[3]	132	72
Rang	1	2
Zusatzpunkte[4]	0	42,5
Minuspunkte[5]	-52,5	0
Gesamt	79,5	114,5
Rang neu	2	1

[1] Die Ergebnisse sind ausgedrückt in Form von Platzierungen innerhalb einer Prüfung (2 = 2. Platz usw.)

[2] a steht für Nichtbeenden des Parcours wegen Ausscheiden oder Aufgabe

[3] Die Anzahl der jeweiligen zu vergebenen Punkte pro Platzierung ist angelehnt an das Reglement der FN

[4] Jeweils die Hälfte der Punkte des letztplatzierten Pferdes (für unplatzierte Leistungen, die zu den besten 50% einer Prüfung gehören)

[5] Abzug der halben Punktzahl des letztplatzierten Pferdes für Nichtbeenden des Parcours bei entgegengesetzter Punkteverteilung (z.B. kostet ein Scheitern im Parcours mit dem höchsten Schwierigkeitsgrad die halbe Punktzahl des Letztplatzierten im Parcours mit dem geringsten Schwierigkeitsgrad)

Im Ergebnis bestätigen die Beispiele die These, dass eine Gesamtbetrachtung aller Starts und eine differenziertere Vergabe von RP grundsätzlich deutliche Auswirkungen auf das Ranking und damit auf die Aussagekraft der Leistungsbeurteilung haben können. Am Beispiel von Pferd C, welches die Leistungsgrenze offensichtlich mit der Klasse L erreicht, kann die Vorteilhaftigkeit der zusätzlichen Ermöglichung von „individuellen Rankings" gezeigt werden.

Auf Basis der vorangegangenen Erwägungen sollen in dem folgenden Experiment vier Hypothesen überprüft werden:

(1) Die Bildung von Durchschnitten und die Vergabe von Zusatz- und Minuspunkten unter Berücksichtigung nichtplatzierter Turnierergebnisse führt zu Veränderungen in der nationalen und internationalen Rangierung von Sport-Springpferden.

(2) Die Veränderungen in der Rangierung fallen im Ranking der FN stärker aus als bei der FEI.

(3) Die beiden betrachteten Maßnahmen (Durchschnittswerte bzw. Vergabe von Zusatz-/Minuspunkten) haben unterschiedlich starke Auswirkungen auf die Rangierungen.

(4) Durch die genannten Modifikationen kann eine höhere Validität des Rankings von Sport-Springpferden erreicht werden.

Systemmodellierung

Ein neues Rankingsystem, das die geforderte Gesamtbetrachtung ermöglicht, muss neben der Einhaltung der Gütekriterien auch folgende Mindestanforderungen erfüllen: Datenverfügbarkeit, Praktikabilität und Akzeptanz bei den Zielgruppen. Ein einfaches durchschnittsbildendes System berücksichtigt alle Ergebnisse und ist problemlos für große Starterfelder, wie sie der Reitsport bietet, technisch umsetzbar. Die zusätzlich erforderlichen Daten (Anzahl der Starts und vollständige Prüfungsergebnisse) liegen der FN heute bereits vor. Alternativ könnte eine Verarbeitung aller Ergebnisse auch durch ein akkumulierendes System erreicht werden, wenn die abgestufte Punktevergabe für alle Teilnehmer an Prüfungen fortgesetzt wird. Auch hier wäre die Umsetzung aus bestehenden Informationen zu bewerkstelligen. Die Leistungskonstanz eines Pferdes hingegen kann nur mittels eines Durchschnittsverfahrens abgebildet werden. Aus diesen Gründen wird im Folgenden ein Übergang vom derzeit angewendeten kumulativen Verfahren auf ein Durchschnittsverfahren mit realen Erfolgsdaten untersucht.

Für die Berechnung des Durchschnittswertes „RP pro Start" unter Berücksichtigung von Zusatz- und Minuspunkten sind folgende Daten zu dokumentieren: Name und Datum des jeweiligen Turniers, Nummer, Schwierigkeitsklasse und Ergebnisliste der einzelnen Prüfungen (mit Pferdenamen), Gewinnsummen, ausgeschiedene Pferde und Anzahl der Teilnehmer und Platzierungen. Damit die Ergebnisse aussagekräftig sind und um „strategisches Handeln" im Hinblick auf die Ranglistenplatzierung zu unterbinden, ist es erforderlich, eine Mindestanzahl von

Starts (z.B. 10) zu definieren. Dies ist bei den Sportarten, die bereits ein „Averaging System" verwenden, gängige Praxis (Stefani, 2011). Bei der Festlegung des Betrachtungszeitraums wird ein gleitendes Zeitintervall der letzten 12 Monate gewählt. Die Betrachtung einer starren Periode hätte zur Folge, dass zu Beginn einer Periode keine Werte vorhanden wären. Außerdem wird durch den gleitenden Betrachtungszeitraum die Aktualität des Ranking verbessert.

Als weitere Komponente werden nun Zusatz- und Minuspunkte analog zum oben diskutierten Beispiel 2 eingeführt. Diese Abgrenzung ist zu einem gewissen Grad willkürlich gewählt; wahlweise könnte etwa die reguläre Punktevergabe von einem Drittel auf die Hälfte ausgeweitet werden. Diese Modifikation, bei dem diverse Varianten denkbar sind, erfordert eine intensive Diskussion im Kreis der Zielgruppen des Rankings. Die Vergabe von Minuspunkten in der Bewertung von unabhängigen Sportarten könnte zu Motivations- und Akzeptanzproblemen führen.

Jedes Pferd i nimmt im Betrachtungszeitraum an einer gewissen Anzahl von Prüfungen teil. M beschreibt im Folgenden die Menge aller Pferde $i = 1, ..., I$. n steht für die festgelegte Mindestanzahl an Starts. $M^{\geq n}$ beschreibt im Folgenden die Menge aller Pferde, die mindestens an n Prüfungen teilgenommen haben. R beschreibt die Menge aller Prüfungen $j = 1, ..., J$ im Betrachtungszeitraum. Die Teilmenge R_i ist die Menge aller Prüfungen, an denen Pferd i teilgenommen hat. $P(i, j)$ ist die Punktefunktion, die die von Pferd i in Prüfung j erreichten RP nach gegenwertigem System angibt. Die Anzahl der relevanten Prüfungen sowie die erhaltenen RP, die sich aus der jeweiligen Wertigkeit der Prüfungen ergeben, werden gemäß den Reglements der FEI und FN (FEI, 2013; FN, 2013a) erfasst. Der Wert x_i stellt schließlich die Summe der RP für Pferd i in der betrachteten Periode dar. Für alle Formeln gilt die Bedingung: $i \in M^{\geq n}$.

Formel 1   bildet die kumulierten RP x_i a

$$x_i := \sum_{j \in R_i} P(i, j)$$

Formel 2   Bildung des Durchschnittswerts RP pro Start $\bar{x}_i$

$$\bar{x}_i := \frac{1}{|R_i|} \sum_{j \in R_i} P(i, j)$$

Wobei $|R_i|$ die Anzahl der Prüfungen in der Menge R_i darstellt. Um die Vergabe weiterer RP zu berücksichtigen, wird die modifizierte Punktefunktion $\tilde{P}_j(i,j)$ eingeführt. Die Zusatz- und Minuspunkte werden in Abhängigkeit vom letztplatzierten Pferd einer Prüfung l_j vergeben.

Es gilt

$$\tilde{P}(i,j) = \begin{cases} P(i,j), \text{ wenn } P(i,j) > 0, \\ 1/2\, P(l_j,j), \text{ wenn } P(i,j) = 0 \text{ und Pferd i in Prüfung j in erster Hälfte,} \\ -c(K_j), \text{ wenn } P(i,j) = 0 \text{ und Pferd i in Prüfung j ausgeschieden,} \\ 0, \text{ sonst.} \end{cases}$$

Dabei entspricht $c(K_j)$ einer von der Schwierigkeitsklasse K der Prüfung j abhängigen Konstante (siehe unten).

Formel 3   modifizierte Gesamtpunktzahl $x_i^{\pm}$

$$x_i^{\pm} := \sum_{j \in R_i} \tilde{P}(i,j)$$

Formel 4   Bildung des Durchschnittswerts RP neu pro Start $\overline{x_i^{\pm}}$

$$\overline{x_i^{\pm}} := \frac{1}{|R_i|} \sum_{j \in R_i} \tilde{P}(i,j)$$

Durch schärfere Abgrenzungen der Mengen ließen sich speziellere Rankings erstellen. So könnte die Menge der Pferde M durch weitere Indizes eingeschränkt werden, z.B. auf ein Zuchtgebiet oder ein Geburtsjahr. Die Menge der Prüfungen R könnte ebenfalls weiter eingeschränkt werden. Für ein Herunterbrechen der Prüfungen auf die einzelnen Schwierigkeitsklassen würde sich R_i wie folgt gestalten: R_i^K sei mit $K \in \{AA, A, B, C, D, E, F, R\}$ die Menge aller Prüfungen, an denen Pferd i im Betrachtungszeitraum teilgenommen hat und die der Schwierigkeitsklasse K entsprechen. In diesem Fall kann $c(K_j) = 1/2\, P(l_j,j)$ gesetzt werden. Im Allgemeinen muss jedoch gewährleistet sein, dass bei einer höheren Schwierigkeitsklasse

Ausscheiden nicht schwerer bestraft wird als Scheitern bei einer leichten Prüfung. Wir wählen daher bei der FEI $c(K)$, mit $K \in \{AA, A, B, C, D, E, F, R\}$, gemäß Tab. 5.

Tab. 5. Vergabe der Minuspunkte in den Prüfungskategorien nach FEI

AA	A	B	C	D	E	F	R
2,5	2,5	5	5	10	20	2,5	5

FEI Internationale Reiterliche Vereinigung

Im Fall der FN entsprechen die Minuspunkte jeweils der Hälfte der Punkte des Letztplatzierten bei entgegengesetzter Wertigkeit der Prüfungen. So wird z.B. ein Scheitern im CSIO3/4/5* mit 15 Minuspunkten bestraft.

Im folgenden Abschnitt soll nun durch einen Test mit Realdaten die Auswirkung der Ergänzungen auf die Rangfolge der Pferde überprüft werden.

Stichprobe

Für das Rankingsystem der FN ist die Überprüfung aller in Wettkämpfen erbrachten Leistungen der Pferde nicht möglich. Ausschlaggebend hierfür ist die mangelnde Informationstiefe der FN-Datenbank, Informationen über nicht platzierte und ausgeschiedene Teilnehmer werden nicht veröffentlicht. Aus diesem Grund wurde für einen Test mit Realdaten die FEI-Datenbank als Grundlage gewählt (FEI, 2012a). Die Stichprobe umfasst die fünfzig besten Pferd-Reiter-Kombinationen des Jahres 2012. Sie stellt im Sinne des vorgestellten Modells die Menge M dar. Da zum Zeitpunkt der Untersuchung kein reines Pferderanking angeboten wurde, fehlen für zwei Pferde Erfolge, die sie im gleichen Zeitraum mit einem anderen Reiter erzielt haben. Die möglicherweise geringfügig andere Reihenfolge im Spitzenfeld bedeutet aber keine Einschränkung in der Eignung der Datenmenge für diesen Test. Die Ergebnisse wurden schrittweise erarbeitet, um die Effekte der einzelnen Erweiterungen zu bewerten.

Zunächst galt es, für jedes der fünfzig Pferde die Anzahl der Starts, das jeweilige Abschneiden pro Prüfung, die erreichten RP und die Anzahl der Teilnehmer pro Prüfung aus unterschiedlichen Tabellen zusammenzustellen. Den Pferden, die in der ersten Hälfte einer Ergebnisliste standen, aber nicht platziert waren, wurde entsprechend obiger Formel die Hälfte der RP des letztplatzierten Pferdes zugewiesen. Für das Ausscheiden eines Pferdes wurden analog

RP in Höhe der Hälfte des Letztplatzierten bei entgegengesetzter Punkteverteilung abgezogen (ein Scheitern in der schwersten Prüfung kostet die wenigsten Minuspunkte usw.).

Datenanalyse und Ergebnisse

Datenanalyse

Die Anzahl der Starts musste bereinigt werden, da es Prüfungen im FEI-Kalender gibt, bei denen keine RP vergeben werden. Hierzu zählen z.B. Prüfungen, die einer besonderen Einladung bedürfen und somit nicht allen Reitern zugänglich sind. Unter Berücksichtigung der Zusatz- und Minuspunkte wurde eine neue Gesamtpunktzahl für das Jahr 2012 gebildet.

Um zu überprüfen, ob sich die Auswirkungen der Erweiterung grundsätzlich auch auf das Rankingsystem der FN übertragen lassen, wurden analog zum oben beschriebenen Vorgehen theoretische RP nach dem Regelwerk der FN ermittelt. Die Unterschiede bei der Bewertung von Ergebnissen zwischen FEI und FN (Tab. 1) wurden beim Übertrag der RP auf das System der FN berücksichtigt. Wegen der nicht trennscharfen Definition der Anzahl zu platzierender Pferde bei der FN wurden analog zum System der FEI in jeder Prüfung RP für die ersten 16 Pferde vergeben. Da die heute abweichenden FN-Schwierigkeitsklassen auf Basis der Preisgelder zu ermitteln sind, war für Turniere außerhalb des Euroraumes eine Umrechnung gemäß den EZB-Referenzkursen erforderlich (Deutsche Bundesbank, 2013).

Zur Auswertung der Daten wurden zunächst Rangfolgen für die einzelnen Teilschritte gebildet. Ausgangslage bildet die Rangliste A der Gesamtpunkte der fünfzig Top-Springpferde (Formel 1). In einem zweiten Schritt setzte man diesen Wert in das Verhältnis zu den bereinigten Starts und bildete eine neue Rangfolge B (Formel 2). Anschließend wurden die Zusatz- und Minuspunkte zu den „alten" Gesamtpunkten hinzu addiert, um die Auswirkungen für sich genommen zu ermitteln (Rangliste C, Formel 3). Die Division dieses Wertes durch die Anzahl der Starts ergab Rangliste D, das Gesamtergebnis beider Maßnahmen (Formel 4).

Die Überprüfung von Hypothesen 1-4 erfolgte über graphische Darstellungen, Korrelationsanalysen sowie statistische Testverfahren (im Wesentlichen t-Tests für verbundene Messungen). Für Letztere wurde ein Signifikanzniveau von 0,01 festgelegt. Weitere Details finden sich in den folgenden Abschnitten.

Deskriptive Ergebnisse

Die Abbildungen 1 und 2 zeigen einen paarweisen Vergleich der einzelnen Formeln 1-4, welche zur Berechnung der RP und schließlich zur Bildung der Rangfolgen A-D verwendet wurden. Hier sind jeweils die RP der betrachteten Pferd/Reiter-Kombinationen für je zwei Berechnungsformeln gegeneinander abgetragen (Scatter Plots). Es zeigt sich, dass die Durchschnittsbildung (B und D) die von der FEI (Abb. 1) und FN (Abb. 2) publizierten Rankings maßgeblich verändert. Der Effekt ist weit größer als die Auswirkungen der Addition von Zusatz- und Minuspunkten; bei letzteren (A vs. C und B vs. D) liegen die einzelnen RP fast vollständig auf einer monoton wachsenden Kurve (siehe insbesondere Abb. 1), was bedeutet, dass sich sehr ähnliche – wenn auch nicht gleiche – Rankings A und C bzw. B und D ergeben. Allerdings ist anzunehmen, dass das Ranking von Pferden im Breitensport stärker von Zusatz- und Minuspunkten beeinflusst würde. Spitzenpferde werden zumeist von professionellen Reitern vorgestellt, sodass Minuspunkte durch ein Ausscheiden aus der Prüfung vergleichsweise selten vergeben werden.

Tab. 6 zeigt die Korrelationen der einzelnen Rankings. Hier sind die beiden üblichen rangbasierten Korrelationsmaße (nach Spearman bzw. Kendall) angegeben. Es zeigt sich das gleiche Bild wie in Abb. 1 und 2. Die Korrelationen der ursprünglichen Rankings zur Rangliste D, in der sowohl die Durchschnittsbildung als auch die Vergabe der Zusatz- und Minuspunkte berücksichtigt wird, sind gering. Demzufolge nimmt die Berücksichtigung beider Neuerungen merklichen Einfluss auf die herkömmlichen Rangierungen. Allerdings sind diese Unterschiede in erster Linie auf die Durchschnittsbildung zurückzuführen, wie sich aus den extrem hohen Korrelationen von Ranking A und C sowie B und D ablesen lässt. Diese Aussagen gelten im Wesentlichen sowohl für die Punktevergabe gemäß FEI als auch FN. Für Letztere machen sich jedoch auch Auswirkungen von Zusatz- und Minuspunkten noch stärker bemerkbar (Abb. 2, Mitte).

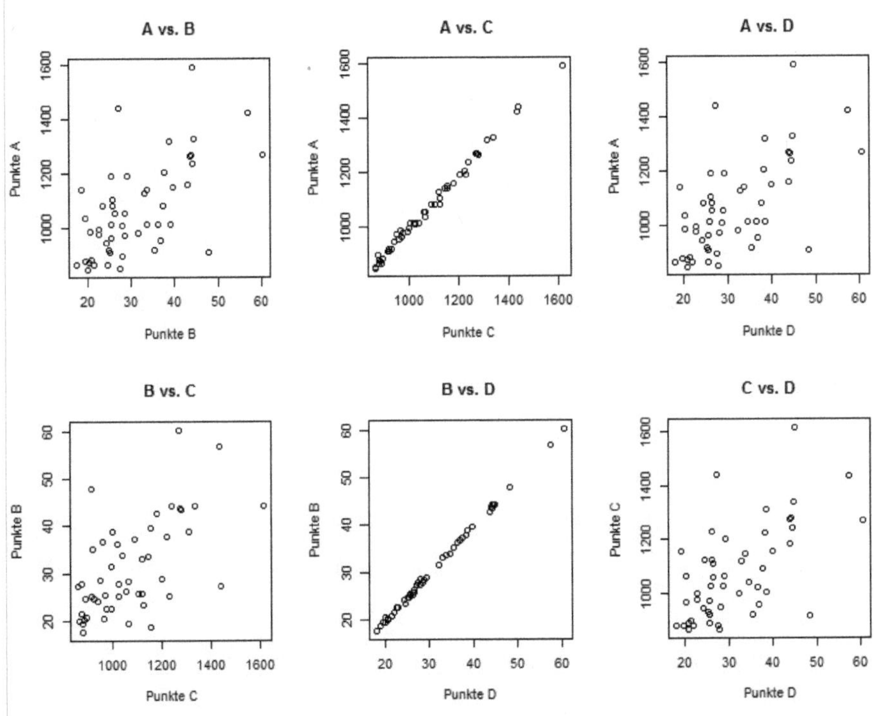

Abb. 1. Paarweiser Vergleich der Punkte A (Formel 1 – Kumulation), Punkte B (Formel 2 – Durchschnittsbildung unter Berücksichtigung von Formel 1), Punkte C (Formel 3 – Berücksichtigung von Zusatz- und Minuspunkten) und Punkte D (Durchschnittsbildung unter Berücksichtigung von Formel 3); jeder Punkt steht dabei jeweils für eine Pferd-Reiter-Kombination. Rohpunkte gemäß FEI

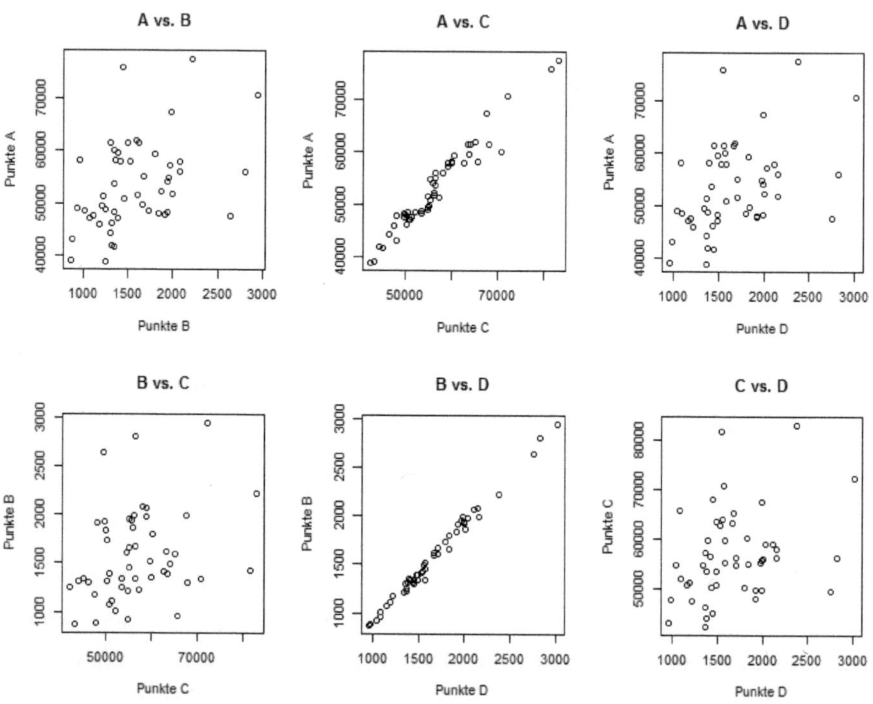

Abb. 2. Paarweiser Vergleich der Punkte A (Formel 1 – Kumulation), Punkte B (Formel 2 – Durchschnittsbildung unter Berücksichtigung von Formel 1), Punkte C (Formel 3 – Berücksichtigung von Zusatz- und Minuspunkten) und Punkte D (Formel 4 – Durchschnittsbildung unter Berücksichtigung von Formel 3); jeder Punkt steht dabei jeweils für eine Pferd-Reiter-Kombination. Rohpunkte gemäß FN

Tab. 6. Korrelationen der einzelnen Rankings

FEI	Spearman			Kendalls τ		
	A	**B**	**C**	**A**	**B**	**C**
B	0,606			0,452		
C	0,994	0,577		0,951	0,426	
D	0,609	0,998	0,583	0,455	0,974	0,433
FN	Spearman			Kendalls τ		
	A	**B**	**C**	**A**	**B**	**C**
B	0,457			0,311		
C	0,977	0,343		0,881	0,221	
D	0,445	0,987	0,342	0,296	0,927	0,220

FN Deutsche Reiterliche Vereinigung, *FEI* Internationale Reiterliche Vereinigung

Prüfung der Hypothesen

In Abb. 1 und 2 sowie Tab. 6 werden die Veränderungen, die sich aus den einzelnen Rankings ergeben, veranschaulicht und quantifiziert. Ein statistischer Test auf Assoziation (d.h. Korrelation ungleich Null) ist hier inhaltlich nicht sinnvoll, formale statistische Tests zur Überprüfung von Hypothese 1 sind nicht erforderlich. Die Hypothese, dass Unterschiede zwischen den Rankings bestehen (d.h. Korrelation < 1) ist trivialerweise erfüllt.[30] Im Folgenden sollen die Hypothesen 2 und 3 näher untersucht werden.

Detaillierte Punkte- und Ranglisten sind in Tab. 11 und 12 im Anhang gegeben. Diese basieren auf der Rangliste der besten Pferd-Reiter-Kombinationen des Jahres 2012 aus den Datenbanken der FEI und der FN. Die in der Rangierung der FEI aufgenommenen Pferde haben an 19 bis 61 Prüfungen teilgenommen. Das auf Platz 1 liegende Pferd „Big Star" hat 1.593 Punkte, der letztplatzierte „Allerdings" hat im gleichen Zeitraum 845 Punkte angesammelt. Der Spitzenreiter „Big Star" verschlechtert sich aufgrund des Teilens durch die bereinigten Starts um 4 Plätze. In Rangliste B liegt er mit 44,25 Punkten pro Start auf Rang 5. Nach Berücksichtigung der Zusatz- und Minuspunkte steigt die Gesamtpunktezahl des Pferdes von 1.593 auf 1.615,5, was in Rangliste C unverändert Platz 1 bedeutet. Der Durchschnitt unter Berücksichtigung der neuen Gesamtpunkte ergibt eine Verschlechterung um drei Plätze (Rangliste D). Die Vergabe der Zusatzpunkte hat zwar im Fall von „Big Star" für sich genommen keine Auswirkungen auf den Rang, in der Gesamtsicht bewirken die zusätzlichen Punkte aber eine

---

[30] Eine (zu Grunde liegende/wahre) Korrelation von 1 entspricht zwingend (d.h. mit Wahrscheinlichkeit 1) einer Übereinstimmung der Rankings. Dies ist hier offensichtlich nicht gegeben.

leichte Kompensation. Das auf dem letzten Rang liegende Pferd „Allerdings" verbessert sich durch die Durchschnittspunkte um vier Plätze. Die Addition der Zusatzpunkte bewirkt eine Verbesserung um einen Platz. Eine erneute Durchschnittsbildung bewirkt eine Verbesserung um 6 Plätze. In diesem Fall haben die Zusatzpunkte eine verstärkende Wirkung.

Tab. 7 zeigt eine Zusammenfassung der gebildeten Differenzen der Ränge aus Tab. 11 und 12 (Auswirkungen der Maßnahmen). Für alle drei Differenzen werden die maximalen Rangveränderungen angezeigt. Der Höchstwert beschreibt die größte Verbesserung der Rangposition, der niedrigste Wert die größte Verschlechterung. Die absoluten Veränderungen sind in vier Kategorien aufgeteilt und zeigen die Häufigkeit der Rangveränderungen in der jeweiligen Rubrik. In der letzten Spalte ist der Mittelwert der (absoluten) Rangveränderungen dargestellt.

Tab. 7. Zusammenfassung der Rangveränderungen

Differenz I – Veränderung des Rangs aus Durchschnittsbildung (ceteris paribus)									
	höchster Wert		niedrigster Wert		absolute Veränderungen				
	Rang-änderung	Name	Rang-änderung	Name	0-10 Plätze	11-20 Plätze	21-30 Plätze	31-40 Plätze	Mittelwert
FEI	37	Carlina	-34	Oz de Breve	32	10	6	2	9,38
FN	37	Carlina	-34	Oz de Breve	22	20	6	2	12,44

Differenz II – Veränderung des Rangs aus Zusatz- und Minuspunkten (ceteris paribus)									
	höchster Wert		niedrigster Wert		absolute Veränderungen				
	Rang-änderung	Name	Rang-änderung	Name	0-2 Plätze	3-4 Plätze	5-6 Plätze	7-8 Plätze	Mittelwert
FEI	3	Lennox	-6	Carlo 273	47	2	1	0	1,12
FN	8	Ornella Mail HDC	-6	Simon	29	12	8	1	2,44

Differenz III – Veränderung des Rangs durch beide Maßnahmen								
	höchster Wert		niedrigster Wert		absolute Veränderungen				
	Rang-änderung	Name	Rang-änderung	Name	0-10 Plätze	11-20 Plätze	21-30 Plätze	31-40 Plätze	Mittelwert
FEI	37	Carlina	-34	Oz de Breve	34	9	5	2	9,38
FN	37	Carlina	-34	Oz de Breve	21	20	7	2	12,76

FN Deutsche Reiterliche Vereinigung, *FEI* Internationale Reiterliche Vereinigung

Differenz I zeigt ceteris paribus die Veränderung des Rangs anhand einer Durchschnittsbildung. Das Pferd mit der höchsten positiven Abweichung, „Carlina", verbessert sich bei der Durchschnittsbildung um 37 Plätze. Wie in Tab. 11 zu sehen ist, startet „Carlina" mit 910 Gesamtpunkten und verbessert sich von Rang 40 auf Rang 3. Hierfür ausschlaggebend ist die mit 19 geringste Anzahl an Starts in der Stichprobe. Das Pferd „Oz de Breve" verschlechtert sich hingegen um 34 Plätze. Mit 61 Prüfungen im Jahr 2012 ist dieses Pferd am häufigsten gestartet und fällt bei den Punkten pro Start von Rang 15 auf Rang 49 zurück. Auf das System der FN übertragen sind die höchsten Abweichungen bei denselben Pferden zu finden. Diese Maximalwerte verdeutlichen wiederum die Auswirkungen einer Gesamtbetrachtung aller Leistungen der Pferde durch das Bilden eines Durchschnitts. Die absoluten Veränderungen zeigen auf, dass bei 36% der Pferde (18) auf Seiten der FEI und 56% der Pferde (28) bei der FN eine Veränderung um mehr als 10 Plätze stattfindet. Der Mittelwert der absoluten Veränderung liegt mit 9,38 bei der FEI niedriger als bei der FN (12,44). Dieser Unterschied ist signifikant (paired t-Test mit $t(49) = 4{,}0168$, $p < 0{,}001$).

Bei Betrachtung von Differenz II wird wiederum ersichtlich, dass die Berücksichtigung von Zusatz- und Minuspunkten ceteris paribus deutlich geringere Auswirkungen auf die Veränderung des Rangs hat als das Bilden des Durchschnitts. Die größte Verbesserung liegt bei „Lennox" vor, der sich in der Rangliste C um 3 Plätze verbessert hat. Die höchsten Punktabzüge werden mit 22,5 bei „Billy Congo" vorgenommen (Tab. 11), was eine Verschlechterung um drei Ränge mit sich bringt. Das Pferd „Carlo 273" büßt dagegen durch eine Verschlechterung von nur 15 Punkten 6 Plätze im Ranking ein. Es zeigt sich, dass bei der Interpretation der Ranglisten die Veränderungen eines Pferdes nicht für sich genommen wirken, sondern dass die Bewegungen der übrigen Pferde ebenfalls Einfluss haben. Der Mittelwert von 1,12 bestätigt, dass die Vergabe der Zusatzpunkte eine signifikant geringere Auswirkung (paired t-

Test mit $t(49) = 6{,}7186$, $p < 0{,}001$) auf die Rangfolge hat als die Durchschnittsbildung (mit einem Mittelwert von 9,38). Auf das System der FN übertragen ergeben sich bei Differenz II etwas stärkere Rangveränderungen als bei der FEI (paired t-Test mit $t(49) = 4{,}1499$, $p < .001$). Aber auch hier ist zu erkennen, dass mit einem Mittelwert von 2,44 die Rangveränderung im Vergleich zur Durchschnittsbildung mit 12,44 wesentlich geringer ist (paired t-Test mit $t(49) = 8{,}6164$, $p < 0{,}001$).

In Differenz III wird die Veränderung auf den Rang durch beide Maßnahmen überprüft. Nach Vergabe der Zusatz- und Minuspunkte wird erneut der Durchschnitt „RP pro Start" berechnet. Durch die geringen Auswirkungen der Zusatz- und Minuspunkte sind die Veränderungen gegenüber Differenz I nicht groß. Die Pferde mit den höchsten und niedrigsten Abweichungen bleiben „Carlina" und „Oz de Breve". Die Mittelwerte weisen allenfalls geringe, nicht signifikante (paired t-Test) Abweichungen zu Differenz I auf. Der Übertrag der FEI-Daten auf das FN-Punktesystem zeigt auch bei Differenz III, dass die Veränderungen der Ränge in die gleiche Richtung gehen, jedoch stärker ausgeprägt sind (paired t-Test mit $t(49) = 4{,}1016$, $p < 0{,}001$). Die Abweichungen zwischen den Systemen sind auf die Unterschiede in der Punktevergabe zurückzuführen. Das Rankingsystem der FEI ist durch die grundsätzlich abgestufte Punktevergabe von Platz 1 bis 16 differenzierter als das System der FN. Hier werden die Punkte lediglich zwischen Platz 1 bis 6 differenziert. Alle weiteren Platzierten bekommen die gleichen RP wie Platz 6. Die Differenz der Punktespanne ist bei der FEI größer und das System reagiert weniger stark auf Schwankungen. Die Rangveränderungen sind dadurch geringer (FEI, 2013).

Die Hypothesen 2 und 3 können somit bestätigt werden. Die herbeigeführten Rangveränderungen sind für beide Modifikationen im System der FN größer als in dem der FEI. Zudem bewirkt die Durchschnittsbildung in beiden Systemen stärkere Veränderungen des Rankings als die Vergabe von Zusatz- bzw. Minuspunkten.

Die abschließende Validierung der verschiedenen Rankings (Hypothese 4) ist nicht trivial, denn bisher ist kein Goldstandard verfügbar, der die Leistungsstärke eines Pferdes objektiv wiedergibt. Die untersuchten RP/Rankings sind gerade ein Versuch, die Leistungsstärke zu messen. Der erste Ansatz ist daher, die auf den Ergebnissen aus 2012 basierenden Rankings mit einer neuen „Validierungsstichprobe" zu vergleichen. Zu diesem Zweck ziehen wir die bis dato erzielten Ergebnisse aus 2013 heran und untersuchen den Zusammenhang der verschiedenen Punkte/Rankings aus 2012 mit den entsprechenden aus 2013. Ein starker Zusammenhang wird als Indiz für Validität interpretiert. Tab. 8 zeigt die Korrelationen (nach Spearman und Kendall) zwischen 2012 und 2013 jeweils für die einzelnen Punkte/Rankings (nach FEI).

Tabelle 8. Korrelationen der Ranglistenpunkte aus 2012 und 2013 für die verschiedenen Berechnungsformeln

	Punkte A	Punkte B	Punkte C	Punkte D
Spearman	0,459	0,446	0,453	0,429
Kendalls τ	0,332	0,318	0,330	0,308

Hier zeigt sich, dass die Zusammenhänge zwischen 2012 und 2013 für alle Rankings allenfalls als moderat zu bezeichnen sind und dass kein Ranking als überlegen beschrieben werden kann. Auch hat diese Art der Validierung einen gravierenden konzeptionellen Schwachpunkt. Man denke nur an ein triviales Ranking, welches die Pferde lediglich in alphabetischer Reihenfolge nach ihrem Namen ordnet. Ein solches Ranking wäre für 2012 und 2013 exakt identisch und würde von daher absurderweise als bestmögliches ausgezeichnet.

Als Alternative zu den Rankings selbst bieten sich die für 2013 erhaltenen Preisgelder an. Hierbei werden sowohl die kumulierten Preisgelder pro Pferd als auch für jedes Pferd die durchschnittlichen Werte pro gerittenem Turnier betrachtet. Werden die Rohpunkte nach FEI vergeben, schneidet – gemessen an den Korrelationen – jeweils Punktevergabe C am besten ab (knapp vor A). Werden die Rohpunkte nach FN vergeben, sind die Ergebnisse uneinheitlich (aus Platzgründen wird hier auf die detaillierte Darstellung der Korrelationen verzichtet). Klare Indizien liefert also auch diese Strategie nicht.

Eine weitere, des Öfteren angewandte Validierungsmethode ist, einen spezifischen Wettkampf von herausragender Bedeutung herauszugreifen und die Übereinstimmung von zuvor aufgestelltem Ranking und dem Abschneiden der Sportler bei eben jenem Wettkampf zu untersuchen (Gerhards & Wagner, 2008; West, 2006). Im Bereich Springreiten (oder auch Reitsport allgemein) ist zweifelsohne eines der wichtigsten Turniere – wenn nicht das wichtigste – der CHIO Aachen. Wir betrachten daher die Korrelation von RP (Version A, B, C bzw. D) aus 2012 und das Abschneiden in Aachen 2013 als Indikator für die Validität der entsprechenden Formeln zur Berechnung der RP. Das Abschneiden in Aachen wird dabei sowohl über die aktuell vergebenen RP (Version A) nach FEI und FN quantifiziert als auch über die modifizierte Formel bei Vergabe von Zusatz- und Minuspunkten (Version C). Die sich ergebenden Korrelationen (nach Spearman und Kendall) sind in folgender Tab. 9 dargestellt.

Tabelle 9. Zusammenhang zwischen verschiedenen Formeln zur Berechnung von Ranglistenpunkten für 2012 und dem Abschneiden der Pferde beim CHIO Aachen 2013

Punkte in Aachen 2013 nach aktueller Definition				
Korrelation mit RP 2012		**Punkte A**	**Punkte B**	**Punkte C**	**Punkte D**
FEI	**Spearman**	0,575	0,657	0,575	0,652
	Kendall	0,404	0,476	0,394	0,453
FN	**Spearman**	0,118	0,376	-0,030	0,309
	Kendall	0,061	0,268	-0,037	0,243
Punkte in Aachen 2013 bei Vergabe von Zusatz- und Minuspunkten				
Korrelation mit RP 2012		**Punkte A**	**Punkte B**	**Punkte C**	**Punkte D**
FEI	**Spearman**	0,596	0,666	0,594	0,662
	Kendall	0,420	0,493	0,411	0,469
FN	**Spearman**	0,096	0,345	-0,018	0,270
	Kendall	0,072	0,227	0,000	0,191

FN Deutsche Reiterliche Vereinigung, *FEI* Internationale Reiterliche Vereinigung

Es ist ersichtlich, dass in allen Fällen Punkte B (d.h. Durchschnittsbildung für 2012) die stärkste Assoziation mit den Ergebnissen aus Aachen 2013 aufweisen, (knapp) gefolgt von Punkte D. Diese Erkenntnis legt nahe, dass durch Durchschnittsbildung das wahre Leistungsvermögen eines Pferdes in der Tat besser abgebildet werden kann als durch einfache Kumulation der in einem Jahr erzielten Punkte. Darüber hinaus fällt beim Blick auf Tab. 9 auf, dass der Zusammenhang für die RP nach FEI deutlich größer ist als bei der Vergabe nach FN. Dies stützt die Behauptung, dass das System der FEI eine differenziertere und damit validere Punktevergabe zulässt.

Diskussion

Im Ergebnis hat der Test mit realen Daten nachgewiesen, dass die Verwendung der durchschnittlichen Punkte pro Start, selbst im absoluten Spitzenfeld mit hoher Leistungsdichte, eine erhebliche Auswirkung auf das Ranking der Pferde hat. Das ist im Wesentlichen zurückzuführen auf ein für den gesamten Reitsport typisches Merkmal: Die Anzahl der Starts der einzelnen Pferde innerhalb eines Jahres schwankt erheblich. Während ein kumulatives System Vielstarter bevorteilt, geben Durchschnittswerte eine davon unabhängige Leistungsbeurteilung ab, belohnen aber unter Umständen Reiter, die sich mit ihren Pferden auf wenige Wettkämpfe konzentrieren. Die zusätzliche Vergabe von RP außerhalb der Platzierungen und die Vergabe von Minuspunkten bei Prüfungsabbruch spielt beim Ranking der Spitzengruppen eine eher untergeordnete Rolle. Diese RP werden in niedrigeren Rängen bei weniger erfahrenen oder leistungsschwächeren Reitern und Pferden häufiger verteilt und sorgen insbesondere dort für eine differenziertere Bewertung. Letzteres wäre im Zuge eines vergleichbaren Experiments im Breitensport näher zu prüfen.

Insgesamt konnte nachgewiesen werden, dass die Rangveränderungen im System der FN stärker ausfielen als in dem der FEI. Dies ist wohl zurückzuführen auf die differenziertere Rankingmethode des internationalen Verbands. Vor allem durch die konsequente Berücksichtigung der ersten 16 Teilnehmer einer Prüfung ist das internationale Ranking transparenter und damit aussagekräftiger. Demnach erweist sich die umfassendere Berücksichtigung von Ergebnissen als vorteilhaft. Es ist anzunehmen, dass die eingeschränkte und variierende Erfassung von Ergebnissen nicht nur im Reitsport, sondern auch in jeder anderen Sportart die Qualität eines Rankings schmälert.

Die Ergebnisse der Überprüfung von Hypothese 4 stützen die Vermutung, dass eine differenziertere Rankingmethode die Leistungsfähigkeit eines Pferdes – hier gemessen am Abschneiden beim CHIO Aachen im Folgejahr (2013) – besser wiedergibt als das aktuell verwendete rein kumulative Verfahren.

Für die Umsetzung der beschriebenen Systemerneuerungen sind die Interessen von hauptsächlich fünf Anspruchsgruppen zu erwägen: Reiter, Züchter und Nachfrager auf dem Markt für Sportpferde sowie Turnierveranstalter und der Dachverband. Die beiden Letzteren müssten auch die Umsetzung übernehmen. Die FN müsste ihr herkömmliches Ranking durch das neue ersetzen oder zumindest ergänzen und die auf Basis der Bildung von Durchschnitten ermittelten Rangierungen veröffentlichen. Die Berücksichtigung aller Ergebnisse der ersten 50% einer jeden Prüfung beinhaltet die Archivierung einer größeren Menge an Daten. Auch basiert das neue Ranking auf der Berechnung komplexerer Formeln als jenes, welches ausschließlich die erreichten RP aufaddiert. Demnach wäre mit der Umstellung ein (leicht) erhöhter Arbeitsaufwand für die FN verbunden, zumindest in der Phase der Implementierung. Des Weiteren ist auf die politische Aufwandskomponente der Neuerung für den Reiter-Dachverband zu verweisen: Die Auseinandersetzung mit den Anspruchsgruppen ist ein unabdingbarer Schritt im Management von Change-Prozessen (Rüegg-Stürm, 1998). Die FN ist das Hauptorgan in der Organisation des deutschen Reitsports, dementsprechend würde sie den Wandel nicht nur durchsetzen, sondern auch rechtfertigen müssen. Administrativer Aufwand entsteht zudem bei den Turnierveranstaltern insofern, als dass die Anzahl der zu veröffentlichenden Ergebnisse von höchstens einem Drittel der Starts je Prüfung auf 50% der Starts je Prüfung anwächst.

Angesichts dessen, dass die durchschnittsbildende Methode weniger die Häufigkeit der Turnierstarts honoriert, sondern vielmehr auf die Qualität der einzelnen Turnierteilnahmen eines Pferdes abstellt, würde das neue System möglicherweise zu einem Rückgang der Anmeldungen für Turnierveranstaltungen führen. Während das kumulative Ranking die erfolgreichen „Vielstarter" bevorzugt, veranlasst die Durchschnittsmethode Reiter ausschließlich dann zu weiteren Teilnahmen an Turnieren, wenn sie entweder die vorgegebene Mindestanzahl an Starts noch nicht erreicht haben oder wenn sie ein erfolgreiches, rankingverbesserndes Absolvieren des Parcours erwarten. Dieser Aspekt ist bei der Erwägung der Interessen aller Anspruchsgruppen zu berücksichtigen. Für die Reitsportverbände und Turnierveranstalter bedeutet dies ggf. Verluste von Nenngeldern, gleichzeitig steigt aber die Aussagekraft des Rankings in Bezug auf die erbrachten Leistungen der Turnierpferde. Um ein Pferd günstig im Ranking zu platzieren, ist es sinnvoller, auf weniger Turnieren bessere Ergebnisse in höheren Klassen zu erzielen. Für häufiges Starten in niedrigen Klassen fehlt der Anreiz. Die Sportpferde müssten während der Saison weniger häufig eingesetzt werden, und in der Konsequenz könnte das neue Rankingsystem die Belastung der Pferde – u.U. auch die Häufigkeit von Doping-Fällen – eindämmen (Gille, Hoischen-Taubner & Spiller, 2011). Durch die Kombination mit einer höheren Gewichtung der Qualität einzelner Starts ist anzunehmen, dass das durchschnittsbildende System den deutschen Reit-Turniersport hinsichtlich seiner Pferdefreundlichkeit und Attraktivität für Zuschauer langfristig profitieren ließe.

Den Anspruchsgruppen der Reiter, Züchter und Pferdekäufer ist gemein, dass sie an einer hohen Validität des Instrumentes zur Qualitätsmessung ihrer Pferde oder der Pferde, die sie kaufen oder mit denen sie züchten wollen, interessiert sind. Durch die Bildung von Durchschnittswerten, der Berücksichtigung von unplatzierten Ergebnissen und Minuspunkten ist die Leistungsmessung anhand des neuen Ranking detaillierter und stärker auf den einzelnen Turnierstart fokussiert. Allerdings wird die Leistungskonstanz der Pferde etwas weniger gut abgebildet. Mehrmonatige Pausen wegen Verletzungen oder geringer physischer oder psychischer Belastbarkeit wirken sich in vermindertem Ausmaß auf die Rangierung aus. Dennoch rangieren auch nach dem neuen Messsystem Pferde mit vielen guten Ergebnissen vor Pferden mit einer geringeren Anzahl guter Ergebnisse. Wird das neu entwickelte Ranking mit der Veröffentlichung von Turnierstarthäufigkeiten der Pferde kombiniert bzw. setzt man eine genügend hohe Mindestanzahl an Turnierstarts voraus, ist eine aussagekräftigere Leistungsmessung gegeben. Dies impliziert eine verbesserte Fairness durch das Ranking und schafft überdies mehr Transparenz in den Preisbildungsprozessen auf den Märkten für Sportpferde. Insgesamt wird deutlich, dass der Übergang zur neu entwickelten Rangierungsmethode einen Mehrwert für den deutschen Turniersport generieren würde.

Die Ergebnisse des hier vorgestellten Experiments am Beispiel Springreiten verdeutlichen die praktische Relevanz von aussagekräftigen Leistungsindikatoren. Darüber hinaus konnte eine weitere Datengrundlage für die kritische Hinterfragung rein kumulativer Rankingsysteme im internationalen Leistungssport geschaffen werden, anknüpfend an die o.g. Studie von Jeremic und Radojicic (2010).

Kumulative Rankings werden zumeist in den unabhängigen Sportarten, in denen kein direkter Kontakt zwischen den Kontrahenten erlaubt ist, eingesetzt. Beispiele hierfür sind außer Reiten auch Schwimmen oder Bogenschießen. Weiterhin gehören 67 Sportarten den objektbezogenen Sportarten an und 18 Sportarten zählen zu den Kampfsportarten (siehe für den Einsatz von Rankingmethoden Tab. 10).

Tab 10. Rankingsysteme im internationalen Sport[31] (in Anlehnung an Stefani, 2011)

	Unabhängige Sportarten[1]	Objektbezogene Sportarten[2]	Kampfsportarten[3]	Gesamt
Kumulative Systeme	3	53	28	84
Adjustive Systeme	1	3	10	14
Subjektive Systeme	2	0	0	2
Mit Rankingsystem insgesamt	6	56	38	100
Ohne Rankingsystem	12	18	30	60
Sportarten insgesamt	18	74	68	160

[1] Ein direkter Kontakt zwischen den Kontrahenten ist nicht erlaubt (z.B. Schwimmen, Reiten, Bogenschießen)

[2] Ein indirekter Kontakt zwischen den Kontrahenten ist erlaubt, wenn sie versuchen, ein Objekt zu kontrollieren (z.B. Basketball, Fußball)

[3] Physischer Kontakt zwischen den Kontrahenten findet statt (z.B. Boxen, Ringen)

Es ist anzunehmen, dass zum einen die intuitive und unkomplizierte Umsetzung und zum anderen der gesetzte Anreiz für Sportler, an möglichst vielen Wettkämpfen teilzunehmen, auch

---

[31] In Anlehnung an Stefani (2011)

in weiteren Sportarten neben dem Reiten ausschlaggebend für den häufigen Einsatz kumulativer Rankings sind. Insbesondere vor dem Hintergrund des sportartenübergreifenden Dopingproblems und der Nachfrage nach validen Entscheidungsgrundlagen für Investoren, stellt sich die Frage, ob kumulative Systeme entsprechend der hier dargestellten Vorschläge, modifiziert werden sollten. Die Optimierung von Instrumenten zur Leistungsmessung im Sport leistet einen Beitrag zur Abwehr von Anreizen, die die Gesundheit von Sportlern gefährden. Zudem kann das Image bestimmter Sportarten durch die transparente und faire Leistungsmessung verbessert bzw. aufrechterhalten werden.

Für die weiterführende Forschung ergeben sich vorrangig zwei Aufgaben. Zum einen sind die Auswirkungen der Modifikationen im Rankingsystem auf den Breitensport zu erwägen. Es ist anzunehmen, dass Zusatz- und Minuspunkte durch die geringere Leistungsdichte im nationalen bzw. ländlichen Turniersport das Ranking vergleichsweise stark beeinflussen. Hingegen wird der Effekt der reinen Durchschnittsbildung vermutlich geringer ausfallen als im Spitzenfeld. Letzteres ist dadurch begründet, dass Breitensportler tendenziell an weitaus weniger Turnieren teilnehmen als Profisportler. Die Verschiebung beider Effekte sollte gegeneinander abgewogen und die möglichen positiven sowie negativen Auswirkungen auf den Sport sollten eruiert werden. Ein solches Folgeexperiment ist sinnvoll vor dem Hintergrund, dass breitensportlich eingesetzte Pferde weitaus häufiger auftreten als Pferde für den professionellen Sport. Im Jahr 2012 wurden 80.425 Turnierpferde in Deutschland registriert, von denen nur 2.423 Pferde für den internationalen Sport, damit für den Spitzensport, zugelassen sind (FN, 2012). Damit stellt der Freizeit- und Amateursport einen großen Teilmarkt dar, dessen Akteure von der Art der angewandten Leistungsmessung hinsichtlich ihrer Anreize beeinflusst werden.

Zum anderen ist der Mechanismus für die objektive Leistungsevaluation bei Sportpferden weiterzuentwickeln und fortlaufend zu verbessern. Im Fokus steht dabei die Erhöhung von Validität und Objektivität. Der Einbezug möglichst vieler zuverlässiger Leistungsindikatoren und die Ausdifferenzierung von Rankings erweisen sich hierbei als zielführend (vgl. u.a. Jeremic & Radojicic, 2010). Der Forschung sei empfohlen, weitere dieser Leistungsindikatoren in der Pferdeevaluation zu identifizieren und in Rankingsysteme zu integrieren. Auf diese Weise lassen sich Letztere als Instrument zur Verbesserung der Transparenz und der Fairness im Sport und auf dessen Märkten einsetzen.

Fazit

Durchschnittsbildende Rankings und die differenzierte Punktevergabe für nicht platzierte Ritte ermöglichen eine transparentere Bewertung von Sportpferden. Die Berücksichtigung von Zusatz-/Minuspunkten hat in der Stichprobe der fünfzig Top-Springpferde vergleichsweise geringe Effekte auf das derzeit angewendete kumulative Ranking. Dennoch ist anzunehmen, dass in den niedrigeren Leistungsklassen größere Rangveränderungen herbeigeführt würden. Um den Interessen der Akteure gerecht zu werden, empfiehlt es sich, die Aussagekraft des Rankings hinsichtlich der Leistungsfähigkeit von Pferden durch Modifikationen der Methodik zu erhöhen. Jahreslisten verbesserten zudem die Aktualität, und eine Erweiterung der individuellen Rankings würde das Erscheinen von Pferden aus dem Breiten- und Freizeitsport in den Rangierungen ermöglichen. Somit würde z.B. Pferdekäufern eine zuverlässige Unterstützung bei der Kaufentscheidung geboten. Für die Vergleichbarkeit ist eine internationale Harmonisierung wünschenswert.

Literatur

Arnemann, S. (2003). *Haltung von Sportpferden unter besonderer Berücksichtigung der Leistung*. Doktorarbeit, Hannover: Tierärztliche Hochschule Hannover.

Arnold, D. (2010). *Pferdewirtprüfung [Bd. 2]: Nachhaltige Fütterung*. BoD-Books on Demand.

Chen, K., Ali, M., Veeman, M., Unterschultz, J., & Le, T. (2002). Relative importance rankings for pork attributes by Asian-origin consumers in California: Applying an ordered probit model to a choice-based sample. *Journal of Agricultural and Applied Economics, 34*(1), 67-79.

Deutsche Bundesbank (2013, 28. März). *Devisenkursstatistik*. Zugriff am 02. April 2013 unter http://www.bundesbank.de/Redaktion/DE/Downloads/Statistiken/Aussenwirtschaft/ Devisen_Euro_Referenzkurs/stat_eurefd.pdf?__blob=publicationFile

Dohms-Warnecke, T. (2012, 24/28. November). FN aktuell. *Offizieller Pressedienst der Deutschen Reiterlichen Vereinigung e.V.*, 18-25. Zugriff am 16. März 2013 unter http://www.pferd-aktuell.de/files/2/47/482/FN-aktuell_24_12.pdf

Elo, A. E. (1978). *The rating of chess players past and present*. New York: Arco.

FEI (2012a, 8. Juni). *Rules FEI World Cup^{TM} Jumping. 12th edition.* Zugriff am 10.April 2013 unter http://www.fei.org/fei/regulations/past.

FEI (2012b, 31. Dezember). *FEI Ranking / Standing.* Zugriff am 2. Dezember 2013 unter https://data.fei.org/Ranking/List.aspx

FEI (2012c, 31. Dezember). *FEI Ranking / Standing Search: Jumping – Combination in Jumping-N^0 5.* Zugriff am 12. Januar 2013 unter https://data.fei.org/Ranking/Search.aspx?rankingCode=S_WR_RH

FEI (2013, 1. Februar). Jumping Rules. 24th edition. Zugriff am 10. April 2013 unter: http://www.fei.org/sites/default/files/JumpRules_24thEd_2013_mark-up_updated_1.2.13.pdf

FN (2011). *Jahresbericht 2011 FN / DOKR.* Warendorf: FN-Verlag.

FN (2012). *Jahresbericht 2012 FN / DOKR.* Warendorf: FN-Verlag.

FN (2013a). *Leistungs-Prüfungs-Ordnung 2013. Regelwerk für den deutschen Turniersport.* Warendorf: FN-Verlag.

FN (2013b). *Ranglisten Dressur, Springen, Vielseitigkeit und Fahren.* Zugriff am 10. April 2013 unter http://www.pferd-aktuell.de/ranglisten/ranglisten

FN (2013c). *Ranglistenpunkte.* FN-Verlag: Warendorf.

FN (2013d). *FN Erfolgsdaten Sport und Zucht.* Warendorf: FN-Verlag.

Frisby, W. (1986). Measuring the organizational effectiveness of national sport governing bodies. *Journal of Applied Sport Sciences, 11* (2), 94-99.

Gerhards, J., Mutz, M., & Wagner, G. G. (2012). Keiner kommt an Spanien vorbei: Außer dem Zufall. *DIW-Wochenbericht, 79*(24), 14-20.

Gerhards, J., & Wagner, G. G. (2008). Marktwert gegen Zufall-wer wird Fußball-Europameister?. *Wochenbericht, 75*(24), 326-328.

Gille, C., Hoischen-Taubner, S., & Spiller, A. (2011). Neue Reitsportmotive jenseits des klassischen Turniersports. *Sportwissenschaft, 41*(1), 34-43.

Hammann, P., & Erichson, B. (2000). *Marktforschung* (5. Aufl.). Stuttgart: UTB.

Heasman, J., Dawson, B., Berry, J., & Stewart, G. (2008). Development and validation of a player impact ranking system in Australian football. *International Journal of Performance Analysis in Sport, 8*(3), 156-171.

Heuer, A., & Rubner, O. (2009). Fitness, chance, and myths: an objective view on soccer results. *The European Physical Journal B, 67*(3), 445-458.

Heuer, A., & Rubner, O. (2012a). How Does the Past of a Soccer Match Influence Its Future? Concepts and Statistical Analysis. *PloS one, 7*(11), e47678.

Heuer, A., & Rubner, O. (2012b). Towards the perfect prediction of soccer matches. *arXiv preprint arXiv:1207.4561*.

Heuer, A., Mueller, C., & Rubner, O. (2010). Soccer: Is scoring goals a predictable Poissonian process?. *EPL (Europhysics Letters)*, *89*(3), 38007.

Himme, A. (2006). Gütekriterien der Messung: Reliabilität, Validität und Generalisierbarkeit. In S. Albers, D. Klapper, U. Konradt, A. Walter, & J. Wolf (Hrsg.), *Methodik der empirischen Forschung* (S. 383-400). Wiesbaden: DUV.

Jeremic, V., & Radojicic, Z. (2010). A new approach in the evaluation of team chess championships rankings. *Journal of Quantitative Analysis in Sports*, *6*(3), Article 7.

Kladroba, A. (2005). *Statistische Methoden zur Erstellung und Interpretation von Rankings und Ratings*, Berlin: VWF Verlag für Wissenschaft und Forschung.

Löwe, H. (1988). *Pferdezucht* (6. neubearb. Aufl.). Stuttgart: Eugen Ulmer Verlag.

Mamerow, A. (2010). *Das Pferd ist dein Spiegel: Besser reiten durch mentales Training*. Leipzig: Draksal Verlag.

Mukherjee, S. (2012). Identifying the greatest team and captain—A complex network approach to cricket matches. *Physica A: Statistical Mechanics and its Applications*.

Raab, M., & Philippen, P.B. (2008). Auf der Suche nach Einfachheit in Vorhersagemodellen im Sport – Kommentar zum Beitrag von W. Maennig und C. Wellbrock. *Sportwissenschaft*, *38*(4), 464-471.

Radicchi, F. (2011). Who is the best player ever? A complex network analysis of the history of professional tennis. *PloS one*, *6*(2), e17249.

Rue, H., & Salvesen, O. (2000). Prediction and retrospective analysis of soccer matches in a league. *Journal of the Royal Statistical Society: Series D (The Statistician)*, *49*(3), 399-418.

Rüegg-Stürm, J. (1998). Implikationen einer systemisch-konstruktivistischen "Theory of the firm" für das Management von tiefgreifenden Veränderungsprozessen. *Die Unternehmung* *52*(2), 81-89.

Sire, C., & Redner, S. (2009). Understanding baseball team standings and streaks. *The European Physical Journal B*, *67*(3), 473-481.

Stefani, R. (2011). Methodology of officially recognized international sports rating systems. *Journal of Quantitive Analysis in Sports*, *7*(4).

Stölting, E. (2002). Wissenschaft als Sport: ein soziologischer Blick auf widersprüchliche Mechanismen des Wissenschaftsbetriebes. *die hochschule*, *2*, 58-78.

WBFSH (2013). *All about the rankings*. Zugriff am 10. Februar 2013 unter
http://www.wbfsh.org/GB/Rankings/WBFSH%20rankings/All%20about%20the%20rankin
gs.asp

West, B. T. (2006). A simple and flexible rating method for predicting success in the NCAA
Basketball Tournament: Updated Results from 2007. *Journal of Quantitive Analysis in
Sports*, 4(2), Art. 8.

Anhang

Tab. 11. Zusammenfassung der gesammelten FEI-Daten der Top 50 im Springreiten 2012

Pferd	Bereinigte Starts S	Gesamtpunkte P alt	Rangliste A (P alt)	RP alt pro Start	Rangliste B (P alt/S)	Zusatzpunkte	Minuspunkte	Gesamtpunkte P neu	Rangliste C (P neu)	RP neu pro Start	Rangliste D (P neu/S)	Differenz I (Rang P alt - Rang P lt/S)	Differenz II (Rang P alt - Rang P neu)	Differenz III (Rang P alt - Rang P /S)
BIG STAR	36	1593	1	44,25	5	27,5	-5	1.615,5	1	44,88	4	-4	0	-3
CRISTALLO	53	1443	2	27,23	28	30	-35	1.438,0	2	27,13	28	-26	0	-26
NINO DES BUISSONNETS	25	1422	3	56,88	2	12,5	0	1.434,5	3	57,38	2	1	0	1
LONDON	30	1330	4	44,33	4	7,5	0	1.337,5	4	44,58	5	0	0	-1
UCEKO	34	1318	5	38,76	12	0	-5	1.313,0	5	38,62	12	-7	0	-7
CASALL LA SILLA	29	1270	6	43,79	7	5	0	1.275,0	7	43,97	8	-1	-1	-2
FLEXIBLE	21	1267	7	60,33	1	2,5	0	1.269,5	8	60,45	1	6	-1	6
CEVO ITOT DU CHATEAU	29	1262	8	43,52	8	17,5	0	1.279,5	6	44,12	7	0	2	1
SIMON	28	1238	9	44,21	6	2,5	0	1.240,5	9	44,30	6	3	0	3
TALOUBET Z	32	1205	10	37,66	13	17,5	0	1.222,5	11	38,20	13	-3	-1	-3
EMBASSY II	41	1190	11	29,02	22	17,5	-5	1.202,5	12	29,33	22	-11	-1	-11
CHAMAN	47	1190	11	25,32	34	40	0	1.230,0	10	26,17	31	-23	1	-20
SILVANA HDC	27	1157	13	42,85	9	25	0	1.182,0	13	43,78	9	4	0	4
STERREHOF'S TAMINO	29	1152	14	39,72	10	7,5	-5	1.154,5	15	39,81	10	4	-1	4
CODEX ONE	34	1140	15	33,53	19	5	0	1.145,0	16	33,68	19	-4	-1	-4
OZ DE BREVE	61	1140	15	18,69	49	32,5	-17,5	1.155,0	14	18,93	49	-34	1	-34
VDL BUBALU	34	1127	17	33,15	20	2,5	-10	1.119,5	19	32,93	20	-3	-2	-3
ARGENTO	43	1105	18	25,70	31	20	-2,5	1.122,5	18	26,10	32	-13	0	-14
LUNATIC	29	1082	19	37,31	14	30	-5	1.107,0	20	26,36	30	-11	-1	-11
TRIPPLE X III	42	1082	19	25,76	30	42,5	0	1.124,5	17	24,45	38	-20	2	-19
AD RAHMANNSHOF'S BOGENO	46	1082	19	23,52	39	7,5	0	1.089,5	21	37,57	14	5	-2	5
MYLORD CARTHAGO*HN	37	1055	22	28,51	24	15	-5	1.065,0	23	28,78	23	-2	-1	-1
HELLO SANCTOS	40	1054	23	26,35	29	15	-10	1.059,0	24	26,48	29	-6	-1	-6
VERDI III	53	1038	24	19,58	47	32,5	-5	1.065,5	22	20,10	47	-23	2	-23

VOYEUR	40	1015	25	25,38	33	12,5	0	1.027,5	26	25,69	33	-8	-1	-8
STAR POWER	28	1013	26	36,18	16	7,5	0	1.020,5	28	36,45	16	10	-2	10
CORNET OBOLENSKY	30	1013	26	33,77	18	30	-2,5	1.040,5	25	34,68	18	8	1	8
BLUE LOYD 12	26	1012	28	38,92	11	2,5	-10	1.004,5	29	38,63	11	17	-1	17
VDL GROEP VERDI	36	1007	29	27,97	25	25	-5	1.027,0	27	28,53	24	4	2	5
CHESTERFIELD	44	993	30	22,57	41	15	-10	998,0	30	22,68	40	-11	0	-10
BILLY CONGO	48	988	31	20,58	44	12,5	-35	965,5	34	20,11	46	-13	-3	-15
CEDRIC	31	980	32	31,61	21	20	-2,5	997,5	31	32,18	21	11	1	11
VIKING	43	975	33	22,67	40	5	-5	975,0	32	22,67	41	-7	1	-8
NOUGAT DU VALLET	34	970	34	28,53	23	10	-30	950,0	36	27,94	25	11	-2	9
REGINA Z	38	965	35	25,39	32	5	0	970,0	33	25,53	35	3	2	0
LORENA 111	26	955	36	36,73	15	5	0	960,0	35	36,92	15	21	1	21
ABBERVAIL VAN HET DINGESHOF	39	945	37	24,23	38	7,5	-10	942,5	37	24,17	39	-1	0	-2
BELLA DONNA 66	37	917	38	24,78	36	25	-12,5	929,5	38	25,12	37	2	0	1
PLOT BLUE	26	915	39	35,19	17	5	0	920,0	39	35,38	17	22	0	22
CARLINA	19	910	40	47,89	3	7,5	0	917,5	41	48,29	3	37	-1	37
DERLY CHIN DE MUZE	36	907	41	25,19	35	17,5	-5	919,5	40	25,54	34	6	1	7
CARLO 273	32	893	42	27,91	26	10	-25	878,0	48	27,44	27	16	-6	15
ORNELLA MAIL HDC	42	880	43	20,95	43	20	-2,5	897,5	42	21,37	43	0	1	0
CASTLE FOR-BES MYRTILLE PAULOIS	45	878	44	19,51	48	12,5	-10	880,5	45	19,57	48	-4	-1	-4
OHM DE PONTHUAL	43	870	45	20,23	45	17,5	0	887,5	44	20,64	45	0	1	0
LENNOX	35	863	46	24,66	37	27,5	0	890,5	43	25,44	36	9	3	10
COS I CAN	49	863	46	17,61	50	22,5	-5	880,5	45	17,97	50	-4	1	-4
CASTLEFIELD ECLIPSE	40	862	48	21,55	42	17,5	0	879,5	47	21,99	42	6	1	6
REVEUR DE HURTEBISE H D C	31	850	49	27,42	27	25	-10	865,0	50	27,90	26	22	-1	23
ALLERDINGS	42	845	50	20,12	46	22,5	0	867,5	49	20,65	44	4	1	6

FEI Internationale Reiterliche Vereinigung

Tab. 12. Zusammenfassung der gesammelten FN-Daten der Top 50 im Springreiten 2012

Pferd	Bereinigte Starts S	Gesamtpunkte P alt	Rangliste A (P alt)	RP alt pro Start	Rangliste B (P alt/S)	Zusatzpunkte	Minuspunkte	Gesamtpunkte P neu	Rangliste C (P neu)	RP neu pro Start	Rangliste D (P neu/S)	Differenz I (Rang P alt - Rang P lt/S)	Differenz II (Rang P alt - Rang P neu)	Differenz III (Rang P alt - Rang P /S)
BIG STAR	35	77590	1	2216,86	4	5.550	-15	83125	1	2.375,00	4	-3	0	-3
CRISTALLO	53	76050	2	1434,91	27	5.750	-145	81655	2	1.540,66	27	-24	0	-25
NINO DES BUISS	24	70810	3	2950,42	1	1.425	0	72235	3	3.009,79	1	2	0	2
LONDON	29	57180	17	1971,72	10	1.850	0	59030	16	2.035,52	8	8	1	9
UCEKO	34	67520	4	1985,88	9	0	-15	67505	6	1.985,44	12	-4	-2	-8
CASALL LA SILLA	28	58000	14	2071,43	6	1.000	0	59000	17	2.107,14	7	8	-3	7
FLEXIBLE	20	56110	18	2805,50	2	300	0	56410	21	2.820,50	2	16	-3	16
CEVO ITOT DU	28	52300	24	1867,86	15	3.750	0	56050	24	2.001,79	9	10	0	15
SIMON	28	54760	21	1955,71	11	425	0	55185	27	1.970,89	13	11	-6	8
TALOUBET Z	32	51500	26	1609,38	21	3.175	0	54675	31	1.708,59	20	5	-5	6
EMBASSY II	41	61410	7	1497,80	24	2.600	-15	63995	9	1.560,85	26	-17	-2	-19
CHAMAN	47	61400	8	1306,38	36	6.600	0	68000	5	1.446,81	33	-29	3	-25
SILVANA HDC	26	51750	25	1990,38	8	4.425	0	56175	23	2.160,58	5	18	2	20
STERREHOF'S	28	54150	22	1933,93	12	1.600	-15	55735	25	1.990,54	10	11	-3	12
CODEX ONE	33	55220	20	1673,33	18	1.175	0	56395	22	1.708,94	19	2	-2	1
OZ DE BREVE	61	58110	13	952,62	47	7.675	-60	65725	7	1.077,46	47	-34	6	-34
VDL BUBALU	33	59380	11	1799,39	17	1.000	-45	60335	13	1.828,33	17	-5	-2	-6
ARGENTO	43	59580	10	1385,58	28	4.075	-15	63640	10	1.480,00	31	-18	0	-21
LUNATIC	28	57900	15	2067,86	7	4.850	-30	62720	12	1.529,76	28	-12	3	-13
TRIPPLE X III	41	60200	9	1468,29	25	10.600	0	70800	4	1.573,33	25	-25	5	-16
BOGENO	45	48500	34	1077,78	44	1.800	0	50300	38	1.796,43	18	17	-4	16
MYLORD CARTHAGO	38	57850	16	1522,37	23	2.025	-15	59860	15	1.575,26	24	-7	1	-8
HELLO SANCTOS	39	62060	5	1591,28	22	3.150	-35	65175	8	1.671,15	21	-17	-3	-16
VERDI III	53	48950	31	923,58	48	5.925	-15	54860	29	1.035,09	48	-17	2	-17
VOYEUR	40	53810	23	1345,25	31	2.700	0	56510	20	1.412,75	35	-8	3	-12
STAR POWER	27	56030	19	2075,19	5	2.000	0	58030	18	2.149,26	6	14	1	13
CORNET OBOLENSKY	30	49700	29	1656,67	19	5.425	-15	55110	28	1.837,00	16	10	1	13
BLUE LOYD 12	25	47800	38	1912,00	14	300	-45	48055	43	1.922,20	14	25	-5	24

VDL GROEP VERDI	35	50800	28	1451,43	26	4.425	-16	55209	26	1.577,40	23	3	2	5
CHESTERFIELD	44	47090	42	1070,23	45	3.775	-18	50847	36	1.155,61	45	-3	6	-3
BILLY CONGO	48	48570	33	1011,88	46	3.675	-103	52142	34	1.086,29	46	-13	-1	-13
CEDRIC	31	38830	50	1252,58	38	3.425	-15	42240	50	1.362,58	39	11	0	11
VIKING	43	58180	12	1353,02	30	1.750	-15	59915	14	1.393,37	36	-18	-2	-24
NOUGAT DU VALLET	34	44280	45	1302,35	37	2.050	-70	46260	46	1.360,59	41	7	-1	4
REGINA Z	38	61600	6	1621,05	20	1.750	0	63350	11	1.667,11	22	-14	-5	-16
LORENA 111	26	47910	37	1842,69	16	2.000	0	49910	40	1.919,62	15	22	-3	22
ABBERVAIL	39	45900	44	1176,92	42	1.600	-18	47482	45	1.217,49	43	1	-1	1
BELLA DONNA 66	36	48350	35	1343,06	32	5.175	-50	53475	33	1.485,42	30	3	2	5
PLOT BLUE	25	48300	36	1932,00	13	1.425	0	49725	41	1.989,00	11	24	-5	25
CARLINA	18	47550	40	2641,67	3	2.000	0	49550	42	2.752,78	3	37	-2	37
DERLY CHIN DE MUZE	35	46100	43	1317,14	34	4.100	-15	50185	39	1.433,86	34	8	4	9
CARLO 273	32	41960	47	1311,25	35	2.275	-110	44125	48	1.378,91	37	11	-1	10
ORNELLA MAIL HDC	42	51370	27	1223,10	40	5.850	-15	57205	19	1.362,02	40	-14	8	-13
MYRTILLE	45	38960	49	865,78	50	4.175	-45	43090	49	957,56	50	-1	0	-1
OHM DE PONTHUAL	43	47570	39	1106,28	43	3.700	0	51270	35	1.192,33	44	-5	4	-5
LENNOX	34	47100	41	1385,29	29	3.600	0	50700	37	1.491,18	29	12	4	12
COS I CAN	49	43810	46	894,08	49	4.875	-15	47860	44	976,73	49	-3	2	-3
CASTLEFIELD ECLIPSE	39	48750	32	1250,00	39	4.750	0	53500	32	1.371,79	38	-8	0	-6
REVEUR DE H	31	41600	48	1341,94	33	3.425	-35	44990	47	1.451,29	32	15	1	16
ALLERDINGS	41	49550	30	1208,54	41	5.175	0	54725	30	1.334,76	42	-12	0	-12

FN Deutsche Reiterliche Vereinigung

III.2 Success Factors of Equestrian Tourism: Evidence from Germany

Authors: Katia L. Sidali[1], Mattia Eggemann[1], Laura Hartmann[1], Olga Filaretova[1], Andrea C. Doerr[2]

[1]Georg-August-University of Goettingen

[2] University S.ta Maria, Brasil

This article is published in a similar version in 2013 in *Turistica Gen - Giu 2013*, 73–83.

Abstract

Although Germany clearly lies at the top of international ranks in both equestrian sports and horse breeding (D. R. Vereinigung, 2012), research on success factors of equestrian tourism is still lacking. Against this background, this study presents the results of an online survey focused on perception of the own business of 107 owners of horse-riding farms. The findings of a regression analysis conducted both in the whole samples (n=107) as well in a sub-sample of operators specialized into horse activities for hiking customers (n=34) show that different factors have an influence on business success such as customer satisfaction in the whole sample and control practices and positive image in the sub-sample. Accordingly, managerial implications are discussed.

Keywords

Success Factors, Horse Tourism, Consumer Satisfaction

Introduction

The market of equestrian tourism in Germany is moving forward. The number of registered operations with riding activities increased in many regions of Germany in the last years. Accordingly, several activities and projects to develop the equestrian tourism were initiated (Franke, Gonsior, & Tennstedt, 2009). Tennstedt (2008) underlined the fact that although Germany clearly lies at the top of international ranks in both equestrian sports and horse breeding, equestrian tourism is, however, almost unknown. Henceforth, further research in this sector is urgently needed (Tennstedt, 2008).

For instance, in Germany some of the newer demands being placed on equestrian tourism are the growing number of tourists aged 65 or older, changes in traveling behavior, the increase in day and weekend trippers, and an increase in expenditure of almost 25% per person per day (Franke, Gonsior, & Tennstedt, 2009; Freyer, 2009). Furthermore, roughly one million horses living in Germany provide, approximately, 300,000 jobs in the various areas of horse industry such as sport, breeding or husbandry (Tennstedt, 2008). This has led to an exacerbated rivalry as well as increased competition with respect to price and quality. For these reasons, it is necessary to look more closely at the topic of key success factors.

The result is a variety of riding activities that are designed with the common purpose of turning horse sports into an attractive and lucrative business. However, many rural entrepreneurs suffer from limited entrepreneurship capabilities (Veeck, Che, & Veeck, 2006) which leads to high fragmentation and lack of coordination of many operators (Hjalager & Johansen, 2013; Hjalager, 1996). Besides, as stated by Franke, Gonsior, and Tennstedt (2009) the design of riding attractions in a special region is the outcome of a complex process since it implies the successful combination of several governmental, social, economic and environmental factors (Figure 1).

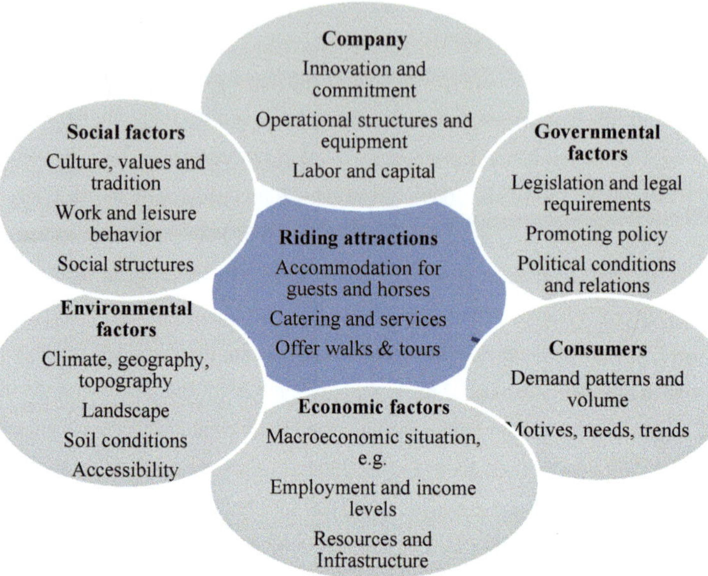

Figure 1. Impact factors of a riding attraction[32]

This study focuses on success factors in equestrian tourism. Specifically, we analyze how rural entrepreneurs, who offer horse riding activities to their guests, interpret the success of their own business.

Success factor research

The area of research that tries to explain 'success' of a business or the reasons why some businesses are more successful than others is called success factor research. In the following, an outline of the different theories on which success factor research is based will be given. Generally, 'success' is considered as being the financial gain, which is thought to be the measurement of economic efficiency, and is not related with business decisions (Tereschenko & Kieneke, 2007). Due to the diversity and complexity of the term 'success', in the field of

---

[32] Inspired by Franke, Gonsior, and Tennstedt (2009).

marketing research however, it is recommended not only to just use economic values such as profit, cash flow or return on investment (ROI) to describe the success of a business, but also to differentiate success with respect to market success and business potential.

Three main approaches have been used in the determination of success. If the research approach focuses on the relationship between success and a business' desired target, this is referred to in the literature as a targeted approach (Tereschenko & Kieneke, 2007). A constituent approach, on the other hand, not only considers economic values with respect to a business's success, it also considers 'success' as being the incentive which a business has to offer its workers in order to achieve a desired behavior pattern and the required contribution to achieve success (Schultze, 2008; Tereschenko & Kieneke, 2007). With a systematic approach, the term 'success' is considered to be the ability of a business to pool scarce and valuable resources as well as to ensure the stability of its business system and to have a successful interaction with its environment (Tereschenko & Kieneke, 2007).

Despite this number of confusing differentiations, all success factors must fulfill three requirements at the company level (Tereschenko & Kieneke, 2007). For instance, (1) they must ensure that the company has a "unique selling position" (i.e. a long-term uniqueness which delimits it from its competitors) (Porter, 1999); (2) they must satisfy the specific requirements of their customers; (3) they must be made up of specific, unique abilities and resources which are difficult to imitate by competitors.

There are numerous relevant approaches which can aid in the explanation of differences in success which come from the field of strategic management. Two central practice-oriented paths within strategic management research are aimed particularly at elucidating 'success'. The first of these is strategy-process research using a planning model while the second path uses a mixture of market-, resource- and competence-oriented approaches (Tereschenko & Kieneke, 2007). Strategy-process research is mainly aimed at answering the question of how the process of strategy formulation and implementation should run so that it positively effects the achievement of objective goals and the business's success (http://strategy.hhl.de). In this context, the basic idea of marketing consists of the consequence orientation of a business on the requirements of its market. According to Tereschenko and Kieneke (2007), the field of marketing research utilises important approaches from the fields of organisational economics and strategic management for its own investigations.

A contrasting approach cited in Tereschenko and Kieneke (2007) is the Dickson's approach. The latter assumes that competitive advantages cannot be achieved through having special characteristics or low costs but through learning processes which arise through experimentation, a continual control of customer satisfaction or through continual benchmarking. The businesses which are more competitive are those which (1) have a pronounced drive for self-improvement, (2) tend to a more pronounced self-reflection than their competitors and (3) implement changes quicker than their competitors (Tereschenko & Kieneke, 2007).

Against this background, it is not surprising that many empirical publications have already tried to explain scientifically the relationship between market orientation and business success, and that marketing is recognized as one of the foundations of business success. Their starting point is the empirical examination of whether the following five individual dimensions associated with management have a positive and significant influence on a business's success: (1) market orientation, (2) employee orientation, (3) environmental and societal orientation, (4) technology and innovation orientation, and (5) production and cost orientation (Fritz, 1993). Other approaches, such as those from Hooley, Lynch, and Shepherd (1990) or Narver and Slater (1990, 1993), displayed a significant influence of market orientation on business success.

One conclusion that has been reached is that businesses should invest in their marketing in good time and not just after the absence of market success as extensive marketing activities normally take time to be effective (Diller, 2006).

The main limitations affecting success in research include basic problems associated with both content and methodology. The problems associated with content are often due to the simplification of causal relationships and their not being adequately revealed. With respect to the methodological problems, often not enough value is placed on a trustworthy and causal interpretation of cross-sectional data, a valid and trustworthy operationalisation and the use of representative sampling as well as unsuitable choices being made with respect to statistical methods (Nicolai & Kieser, 2002).

Success factors in the equestrian sector

Based on literature review, we identified several dimensions that can influence success positively, not only conceived in economic terms but mainly as the self-perception of the own business. The identified dimensions are described as follows.

Business concept: Tereschenko and Kieneke (2007) mentioned service as a potential factor for success. Accordingly, service can also be understood as a product-related resource of a business.

Services: According to Mojen (2009), businesses with higher turnovers provide a higher quality of care for their horses (size of stable, cleanliness of stable etc.). A successful operator of a riding holiday business should offer extra-services such as hire horses to the guests. Strictly connected to this point, the quality of the hired horse and of the horseriding facilities seem to be important dimensions of success.

Customer orientation: Adamer, Hinterhuber, and Kaindl (1993) state that an orientation towards the customers has a positive effect on the success of businesses in every sector. Customers represent resources that can help a business to achieve a competitive advantage (Tereschenko & Kieneke, 2007). Customer orientation is the basis of a long-term stable and economically meaningful relationship with the customer. Adamer, Hinterhuber, and Kaindl (1993) identify customer orientation as assistance of actual and potential customers, flexibility in the fulfillment of customer requests, measures to achieve customer loyalty, etc.

Riding program: The riding program and riding lessons play a large role for a riding holiday business. The riding program is the plan for the riding lessons, while the lessons themselves are defined as the planned imparting of knowledge and skills (Bertelsmann, 1995). Important aspects could be variety in the program, quality, and individual tailoring to the customer.

Marketing advertisement: The main task of marketing or promotion of a business is to capture the attention of individuals and thereby to market goods or services in a targeted manner. The goal is always to increase the distribution, the turnover or the market share (Bertelsmann, 1995; Klapper, Schlichthorst, & Schnell, 2006). As the effect induced by advertising is only detectable by long-term continuous observation, in this article only the following aspects of evaluation of advertising will be considered: the type and variety of advertising and the evaluation of the most effective advertisements or promotional measures (Klapper, Schlichthorst, & Schnell, 2006).

Price for value: The price paid by a customer is often considered relative to the performance of the business perceived by the customer. According to the current literature, the price-performance ratio can be understood as the subjective perception of the appropriateness of the price relative to the quality of the offer.

Certification: Certification of the quality of a riding holiday business positively affects the business success. The certification of riding holiday businesses means a qualification of their services through a seal or label. In Germany, the labeling schemes for horse riding operations with the highest reputation are the DLG (German Agricultural Association) label and the FN (German Equestrian Association) (Ikinger et al., 2013; Wendt, 2013).

Safety: Safety for riders and horses should be guaranteed, especially if the holiday guests are children.

Location of the operation: The location of a business can also be seen as a potential success factor. Riding facilities which are located in the rural borderline of large cities may have an advantage over competitors which are located far away from metropolitan areas. As guests of riding holidays are children in many cases, parents will prefer riding holiday accommodations which are not hours away from their home. Additionally, one can assume that inhabitants of big cities are more likely to decide for holidays in the countryside than inhabitants from rural areas.

Material and Methods

Data were collected online during June and August 2011 in Germany. About 1200 operations, recruited from several online available databanks (e.g. www.reiten-weltweit.de, reiter-ferien.radio101.de) were contacted via mail and invited to answer the questionnaire linked to the homepage of a professional survey provider. The final number of usable data was 107.

Since there are no established scales-of-success factors in the field of horse riding farms available, new ones were developed based on the already discussed literature. The final questionnaire incorporated a set of items regarding the mentioned dimensions influencing success of the business among the sample in general as well as within the subsample of operators specialized into horse activities for hiking customers. The latter are considered as an increasing sub-sector within the equestrian sample (Ikinger et al., 2013; Wendt, 2013).

The final questionnaire for the online survey consisted of two main parts: (1) general information about the respondent's operation (husbandry patterns, type of accommodation, etc.) and (2) the nine dimensions of success of the horse riding business mentioned above.

An exploratory factor analyses (with varimax rotation) were performed to determine which independent factors have possible influence on the success of the business as perceived by the operators. For this purpose, after validating the factors, there were included in two linear regression models. One has been made for the whole sample (n=107) and one for the mentioned subsample (n=34). In both models, a variable of "perception of success" was added as dependent variable (Reisinger & Turner, 1999).

Discussions and Results

A total of 107 available questionnaires were derived from the online survey which corresponds to a response rate of 10.44%. The majority of farms (18.7%) are located in Bavaria, followed by 16.8% located in Lower Saxony, and 11.2% located in Baden-Württemberg. On average, the operations stem from 2000 and 2001 (s.d.=10.42 years) with the oldest operation opened in 1952 and the youngest in 2011. The sample has a high similarity with the national distribution of horse farms (cf.. D. R. Vereinigung, 2012)

We conducted two exploratory factor analyses (EFA) using varimax rotation and factor loading greater than 0.4 criterion on the items representing both the business success and the factors that positively or negatively influence it.

The business success factor, named "Success of enterprise", consisted of three items, namely the farmer's overall satisfaction with his/her operation as well as the evaluation of his/her own success, both in comparison with the sector's average and the main competitors. In the whole sample, the KMO scored 0.562 with a Cronbach's alpha of 0.701 whereas in the sub-sample of operators specialized into horse-activities for hiking customers the KMO and the Cronbach's alpha were respectively 0.562 and 0.673.

The conceptual meaning of each factor is analyzed by observing the items that underlie each of them. Only factors with Cronbach Alpha's values above 0.580 are taken into account.

Factor 1, named "Control (& labeling)" refers to the importance of control practices of the DLG and FN to preserve the quality standards of the operations. Remarkably, in the whole sample both items referring to controls as well as guests' appreciation of such labels merged together. On the contrary, in the subsample only the two items regarding control practices of the DLG and FN built a factor.

Factor 2, called "Satisfaction and image", contains two common items for the whole sample and the subsample. They refer to the satisfaction of the operators regarding the business as well as the positive image that guests have about it. Interestingly, a third item merges to the mentioned items both for the whole sample and for the sub-sample. It is about the willingness to be in the business for the former and an additional item regarding a cozy atmosphere for guests for the latter.

"Customer satisfaction" is the third factor showing the total devotion of operators to the customers even if this implies hiding the own opinion. Table 1 gives an overview of the results of the factor analysis.

Table 1. Summary of exploratory factor analysis for the whole sample (n=107) and the sub-sample (n=34)

Constructs[1]	Overall sample (n = 107)		Riding & Hiking (n = 34)	
	r	Alpha	r	Alpha
Control (& labelling)		0.776		0.856
Item 1	0.827		0.895	
Item 2	0.807		0.821	
Item 3	0.685		--	--
Item 4	0.677		--	--
Satisfaction and image		0.618		0.585
Item 5	0.821		--	
Item 6	--		0.810	
Item 7	0.745		0.599	
Item 8	0.682		0.680	
Customer satisfaction		0.588		0.685
Item 9	0.736		0.822	
Item 10	0.792		0.800	

[1] Five point rating ranging from 1 (totally disagree) to 5 (totally agree).

Item 1 = control made by DLG assure high quality standards; Item 2 = control made by FN assure high quality standards; Item 3 = for guests certification of the riding operation is important; Item 4 = by means of certification the number of guests increases; Item 5= We would start our business over again; Item 6 = We offer our guests a coszy atmosphere.; Item 7 = Satisfied with the own business; Item 8 = Our guests speak frequently about a positive image of our business.; Item 9 = We never show our opinion to the guests.;
Item 10 = We satisfy all needs of our guests.

As mentioned before, in the next step we performed a regression analysis to detect which extrapolated factors have an influence on the own perception of 'success' (Table 2). Since derived success factors are uncorrelated from each other, they could be employed as explanatory variables in the regression analysis whose dependent variable is the factor "Success of enterprise". In the overall sample, the R^2 value indicates that using our model, 16.8% of the variance of the dependent variable cab be explained by independent variables. The Durbin–Watson value of 2.097 is included in the range of 1–3, which implies that errors are independent (Field, 2009). F-value equals 3.456 ($p < 0.001$), which indicates that, overall, the model applied can significantly predict the outcome variable. In the sub-sample the R^2 value scores 85.9% while the Durbin–Watson value is 1.977 and the F-value is 5.598.

Table 2. Path values of regression analysis

Values of path coefficient	Overall sample (n=107)	Riding & Hiking (n = 34)
	Sample mean path coeff. p-value	Sample mean path coeff. p-value
Control -> Success (self perception)	N.S.	0.296
		0.068
Positive image -> Success (self perception)	N.S.	0.593
		0.002
Customer Satisfaction -> Success (self perception)	0.461	N.S.
	0.000	
	R^2-Value	R^2-Value
Success (self perception)	16.8%	85.9%
	Durbin-Watson	Durbin-Watson
	2.097	1.977
	F-value	F-value
	3.456	5.598

In the following, we comment the direction and the magnitude of the path coefficients (Dibbern & Chin, 2005) in order to get some insights into factor invariance among the two samples. For this purpose, several threshold levels are distinguished in the literature corresponding to different levels of model compatibility: from minimal comparability to weak identity. The latter identifies "samples [which] can differ in the level of endorsement of the various items, while everything else—their complete meaning structure, including item reliability—is the same" (Grunert et al., 1997). Based on such considerations, the output of the regression analysis shows that different factors in the two samples have an influence on business success.

In the overall sample, the sign and value of the Beta coefficient (0.461) of the factor customer satisfaction is the main variable which explains the success of the business as perceived by the operators. On the contrary, in the sub sample two different variables play a positive and significant role in explaining business success, namely control practices and positive image (beta coefficients = 0.296 and 0.593, respectively). Since the value of the beta coefficient of the control factor is lower than the factor positive image, it means that coercive practices contribute to the business success with a minor impact. Therefore, practices related to increase the positive image of the business among guests are to be encouraged.

Conclusions

Riding tourism is a growing sector in Germany, but has only recently been recognized as a line of business as it is confirmed by the literature lacuna in this sector. Henceforth, this study aimed at identifying success factors of this increasing niche market from the perspectives of rural operators who offer riding activities to their guests. The findings of the online survey were tested with multivariate statistics.

Interestingly, the results of the literature diverge from those of the online survey. As stated before, among the former, the business concept, the service for customers and horses, the certification of riding holiday businesses using seals, the professional competence of the managers and staff, the countryside for riding and the marketing activities by the riding holiday business have a positive influence on the success of a riding holiday business. But the findings of the multivariate statistics based on the answers of operators running horse-riding farms identified other factors with a positive impact on success. For instance, in order to have a high perception of the own operation as a success one, it is important that the own business is perceived not only as a friendly organization but also as an efficient one. It seems that the perception of success depends on the delicate balance of self-confidence in the own business on the one hand and on the capability to offer guests an affectionate relationship on the other hand. Finally, as stated by Homburg and Stock (2001), success relies on a positive relationship between satisfaction of managers and customer satisfaction, the latter leads to success.

References

Adamer, M., Hinterhuber H., & Kaindl G. (1993). Markt und Weltmarktführer . Was zeichnet diese Betriebe aus? Eine Analyse vor dem Hintergrund der Erfolgsfaktorenforschung. *Der Markt, 32*(124), 6-11.

Bertelsmann (1995). *Neues Lexikon in 10 Bänden.* Bertelsmann Lexikon Verlag, Gütersloh.

D. R. Vereinigung e.V. (2012). *Jahresbericht 2012.* FN-Verlag.

Dibbern, J., & Chin, W.W. (2005). Multi-Group Comparison: Testing a PLS Model on the Sourcing of Application Software Services across Germany and the U.S.A. Using a Permutation Based Algorithm. In F. Bliemel (Hrsg.) *Manual of PLS-path modelling* (pp. 135-160). Stuttgart: Schäffer-Poeschel.

Diller, H. (2006). Probleme der Handhabung von Strukturgleichungsmodellen in der betriebswirtschaftlichen Forschung. *Die Betriebswirtschaft, 66* (6), 611-618.

Field, A. (2009). *Discovering Statistics using SPSS.* Sage Publications Ltd.

Franke, U., Gonsior, I., & Tennstedt, D. (2009). *Tourismus rund ums Pferd, Marktanalyse BTE Tourismusmanagement und FN.* FN Verlag, Warendorf.

Freyer, W. (2009). *Tourismus-Marketing, Marktorientiertes Management im Mikro- und Makrobereich der Tourismuswirtschaft,* 6. Auflage, Oldenbourg Wissenschaftsverlag GmbH, München.

Fritz, W. (1993). Marktorientierte Unternehmensführung und Unternehmenserfolg. *Marketing ZFP – Journal of Research and Mangement, 4*(4).

Grunert, K. G., Brunsø, K., & Bisp, S. (1997). Food-related lifestyle: Development of a cross-culturally valid instrument for market surveillance. In Kahle L., Chiagouris C. (Ed.) (p. 337-354). *Values, lifestyles, and psychographics.* Lawrence Erlbaum Assoc Inc, New Jersey.

Hjalager, A. M. (1996). Agricultural diversification into tourism: Evidence of a European community development programme. *Tourism management, 17*(2), 103-111.

Hjalager, A. M., & Johansen, P. H. (2013). Food tourism in protected areas–sustainability for producers, the environment and tourism?. *Journal of Sustainable Tourism, 21*(3), 417-433.

Homburg, C., & Stock, R. (2001). Der Zusammenhang zwischen Mitarbeiter- und Kundenzufriedenheit: Eine dyadische Analyse. *Zeitschrift für Betriebswirtschaft, 71*(7), 789-806.

Hooley, G. J., Lynch, J. E., & Shepherd, J. (1990). The marketing concept: putting the theory into practice. *European Journal of Marketing, 24*(9), 7-24.

Ikinger, C., Münch, C., Wiegand, K., & Spiller, A. (2013). Reiterleben, Reiterwelten: Zielgruppen zwischen Reitweisen, Motiven und der Liebe zum Pferd. Georg-August-Universität Göttingen, HorseFuturePanel UG, Dietz und Consorten (Eds.). URL: http://www.uni-goettingen.de/de/document/download /1988e74b5e6a7bf92bf38381a71a47f0.pdf/2013-04%20reitsportstudie_screen.pdf. Last accessed: 20 January 2015.

Klapper, D., Schlichthorst, M., & Schnell, C. (2006). Die Analyse langfristiger Werbewirkung. *Marketing ZFP – Journal of Research and Management*, 28(4), 219-235.

Mojen, S. (2009). *Erfolgsfaktoren in der Pensionspferdehaltung. : Kundenzufriedenheit und deren Erfolgswirksamkeit*. Masterarbeit, Georg August Universität, Göttingen.

Narver, J., & Slater, S. (1993). Market orientation and customer service: The implications for business performance. *European Advances in Consumer Research, 1*, 317-321.

Narver, J. C., & Slater, S.F. (1990). The effect of a market orientation on business profitability. *Journal of Marketing, 54*(4), 20-34

Nicolai A., & Kieser A. (2002). Von Konservierungsmaschinen, Nebelkerzen und The Operation called Verstehen. *Die Betriebswirtschaft, 62*, 579-596.

Porter, M. E. (1999). *Wettbewerbsvorteile: Spitzenleistungen erreichen und behaupten*, 6. Auflage, Frankfurt/Main, New York: Campus.

Schultze, D. M. (2008). *Erfolgsfaktoren landwirtschaftlicher Unternehmen mit Marktfruchtbau* (Doctoral dissertation, Martin-Luther-Universität Halle-Wittenberg).

Tennstedt, D. (2008). *Reittourismus in Deutschland: Analyse der gegenwärtigen Situation und zukünftiger Potenziale*. VDM-Verlag Dr. Müller, Saarbrücken.

Tereschenko, O., & Kienecke, T. (2007). *Erfolgsfaktoren*. VDM- Verlag Dr. Müller, Saarbrücken.

Veeck, G., Che, D., & Veeck, A. (2006). America's changing farmscape: A study of agricultural tourism in Michigan. *Professional Geographer, 58*(3), 235-248.

Wendt, J. (2013). *Quantitative Untersuchung zum Image im Reitsport: Ein Vergleich von Eigenbild und Fremdbild*. Masterarbeit, Georg-August-Universität Göttingen.

Summary and Discussion

This dissertation is devoted to attitudes of consumers towards luxury in two agricultural markets, horse sports and foods. Against the background of a postulated change of perceived luxury definitions and motives for luxury consumption (Atsmon et al., 2012; Bellaiche et al., 2012; Meurer, 2012; Yeoman, 2011; Yeoman & McMahon-Beattie, 2010), studies were aimed to reveal how far it has affected the consumer behavior in both agricultural markets. The research results are used to define the target groups for different kinds of luxury marketing and to give recommendations for the design of accordant marketing strategies. Thus, this work helps to fill a vacuum in marketing science: The value change in luxury consumption is a recent phenomenon through which new luxury markets have established (Silverstein, Fiske, & Butman, 2008) and a need for new consumers studies in luxury marketing is created (Meurer & Manniger, 2012). Due to the timeliness of these developments, there is a lack of empirical target group specification in particular luxury markets that are focused on the differentiation between traditional and modern luxury consumption patterns. By investigating German and Chinese horse riders, the added aim of the dissertation was to reveal potential differences in attitudes toward luxury between cultures and to test the validity of modern (Western) consumption motives in a Confucian country. An excursion was used to screen applied marketing instruments or to reveal the success factors for marketing in an environment that is associated with luxury. Altogether, two studies were conducted in German and Chinese horse sports, three studies investigate the German market for luxury foods or upscale foods and two studies represent the excursion.

First, we conducted a survey of 646 **German horse riders or horse lovers** in order to investigate their affinity toward classical luxury products and services. For statistical methods, an explorative factor analysis (principal component analysis) and a three-step cluster analysis were applied. We found that both target groups do exist in this market: people with a traditional luxury understanding, and people who define luxury rather by immaterial values and *luxury experience*. The latter group often refers especially to their horse sports. They perceive it as a luxury to spend time with horses, riding and keeping horses and to have good training conditions in a stable or being advised by well-educated trainers. Contrary to the assumption underlying the study, that horse riders in general can be addressed by luxury marketing strategies, we found another segment that is rather luxury averse. We assume that this group of equestrians has a negative association with the expression of luxury.[33] They reject the link of their sports with a material understanding of luxury.

What we conclude from this study is that luxury marketing strategies on the German horse sports market should reflect the postulated value change in luxury consumption. Only less than one-third of the participants in the survey can be addressed by marketing strategies that are built on a traditional understanding of luxury. Most participants are following immaterial, hedonic or functional motives for luxury consumption. Intangible values, self-fulfillment and quality, which are detached from a stratification effect within a social environment are perceived to be important. The new focus on personally-oriented motives is an outcome of the change toward a modern understanding of luxury (Yeoman, 2011).

The research on the **Chinese market for horse sports** was done by combining quantitative and qualitative data sources. First, a standardized questionnaire for 67 Chinese participants in equestrian activities was quantitatively evaluated. Second, face-to-face interviews with Chinese equestrians, who we randomly met at horse stables and tournaments in Beijing and Shanghai provided quantitative information. Third, a French manufacturer of customized riding saddles was analyzed both quantitatively and qualitatively. The case was chosen in order to provide an example of a company to investigate that operates in the Chinese horse sports market. Finally, the research was built upon general observations in the Chinese equestrian scene. The common use of multiple data collection methods is known to generate positive synergistic effects. Quantitative research can reveal relationships within the data while qualitative methods are used to explain them and build theories (Eisenhardt, 1989*)*. The subordinated research question was if customized handmade leather riding saddles with the highest quality and prices in the market can be successfully marketed among Chinese horse riders. We find that this target group generally has a affinity towards material luxury goods. They have high or very high incomes compared to their compatriots. Further, they are mostly more oriented on Western cultural values like individualism than on Confucian collectivistic values (Hofstede, 1993). Therefore, we show that this target group's main motive for consuming material luxury is not prestige or conspicuousness. They rather value uniqueness, quality and superior craftsmanship. Atsmon, Dixit and Wu (2011), who investigated Chinese luxury consumer in general, find that quality and superior craftsmanship are key factors for luxury purchases.

---

[33] The phenomenon of luxury aversion is for example discussed by Allérès (1993) and Dubois and Laurent (1994).

Compared to German horse riders, we find differences in luxury definitions and luxury motives. Chinese equestrian people seem to be more interested in particular luxury items, they have a higher brand awareness and are following a more material understanding of luxury. This is repeatedly discussed in the literature (Chevalier & Lu, 2010; Lu, 2008; Wong & Ahuvia, 1998; Zhan & He, 2012). But at the same time, our results revealed that luxury values like self-actualization, and functional values which are quality and uniqueness are more significant for the purchase choices of this consumer group than the social aspects of luxury like prestige. In the literature, the Confucian value of collectivism often explains the interests of Chinese people in luxury markets. Acccordingly, the group is more important than the individual and consuming expensive, prestigious luxury goods are used for signaling status in a social environment. Socially-oriented motives, like prestige and conspicuousness, are postulated as preferential for spending money on luxury (Wong & Ahuvia, 1998). This inconsistency with our findings might be explained by the specificities of Chinese equestrian people: their affinity toward a sport that is typical for European nations like Germany, England and France, their high levels of education and high monetary resources. These are assumed as causes why most members of this group have experiences in Western cultures. The interviews with Chinese horse riders, horse owners and stable managers indicated for example that most of the younger target group have studied either in Europe or in the U.S. or spent some time in European stables in order to improve their riding capabilities. To some extent, they have adapted to Western values like individualism and independency, and thus present a target group for luxury marketing, which cannot be exclusively ascribed to socially-related consumption motives. It is without doubt that Confucian values influence the behavior of Chinese luxury consumers. The social dimension of luxury and material values are clearly more significant than in European countries (Heinemann, 2008, Lu, 2008, Schütte & Ciarlante, 1998; Zhan & He, 2012). But in this target group, traditional patterns of consumption have already become less significant. Moreover, there are reasons to assume that the patterns of luxury consumption in China in general will experience a change as well in the future. Chinese equestrians partly present an example of the new generation of Chinese luxury buyers. Our observations show, that they are young, well-educated, open-minded to other cultures and aim to realize their individual dreams. They are not interested in showing-off but they gain personal benefits from consuming quality. According to Degen (2009) and Lu (2008), the Chinese luxury consumers of the future will be the rising generation of well-educated, ambitioned young adults. They aspire to become members of the Chinese elite, but this is done by achievements in their professional careers and living a fulfilled life rather than by showing-off with prestigious consumption. Self-actualization and individualism belong to their previous maxims. Furthermore, Chevalier and Lu (2010) find the Chinese customer group of *luxury intellectuals*, define luxury primarily by analytical, functional and individualistic characteristics. As an example for this target group, they describe a 31-year-old female doctor, who is not interested in following trends and listening to what people think about her. Instead, she relies on facts and just listens to valuable critics on her professional competencies. Her focus is not on fitting into her social environment but on fulfilling herself. The authors show that this group accounts for the greatest part of the population that ultimately adopts luxury products (35.2%). Furthermore, a study by Atsmon et al. (2012) reveals a new preference for understated luxury products among Chinese luxury buyers. More than half of the target group finds that showing-off is in bad taste (51%) and prefers less flashy products (66%). These shares had been grown rapidly; they were 37% and 50% in 2010. The authors further report that Chinese luxury consumption has transformed and they confirm the intensifying relevance of individualistic, self-directed luxury motives.

It cannot presently be stated to what extent the described developments will change the consumption behavior in Chinese luxury markets. It remains questionable if we will observe a general departure from traditional luxury values like prestige, conspicuousness and social stratification. However, the direction of movement toward emphasizing self-directed luxury values cannot, as of recently, be dismissed.

The observed preference of Chinese luxury customers for high quality and functionality, craftsmanship and individuality is combined with a culturally inherent long-term orientation (Hofstede, 1993). According to the latter, the Chinese highly value the stability and liability in social relationships in their culture. Furthermore, harmony, attachment and loyalty are anchored in the Chinese phenomenon of *guanxi*. It refers to stable interpersonal bonds, which present the informal channel for communication. Trust, loyalty and the preference for long-term business relationships are still crucial elements of business practices in China (Lee & Dawes, 2005). As a consequence of the findings, we propose *One-to-one luxury marketing* as a strategy, which should fulfill the needs of the new generation of Chinese luxury customers. It means to market high-end luxury goods with a one-to-one concept, as Peppers and Rogers (1993, 1997) and Peppers, Rogers, and Dorf (1999) describe. The customers would profit from fully individualized products and/or services and a close relationship with the manufacturers.

In **conclusion**, both studies on equestrian people confirm that social luxury values like prestige and conspicuousness have lost significance in the markets for horse sports. Instead, modern intangible, hedonic and personally-oriented luxury values are meanwhile predominant for the explanation of luxury consumption patterns. The study of Chinese equestrian people shows that this target group is generally more expected to buy material luxury, while the motives for consuming luxury goods are individualistic and self-directed. This implicates that a change in luxury consumption patterns has even occurred cross-culturally. Nevertheless, the interpreter of the results of both studies of equestrian athletes should take into account that the representativeness of the samples is restricted. Especially, people with high education are over-represented.

Two further studies in this dissertation investigate the **perceived dimensions of luxury foods**. The results are based on a representative survey of 936 German consumers in the food market. We employed a model inspired by Wiedmann, Hennigs, and Siebels (2007, 2009) and thus tested the validity of a financial, functional, individual, social and the *new* perceived luxury dimension for the case of luxury foods. In the first study, an explorative factor analysis (principal component analysis) reveals the relevant dimensions. These are employed as cluster building variables in a cluster analysis that segments the market. In the second study, a hierarchical partial least squares path model analysis[34] is conducted for testing the causalities between the dimensions and a latent luxury food value. The factors estimated in the first paper on luxury food inspire the choice of the first-order latent construct variables. An increasing segmental interest and willingness to pay by German consumers for food quality, indulgence, as well as health and sustainability concerns motivate the research question for both studies (Kirig & Ruetzler, 2007; Nestlé Deutschland AG, 2012; Nitzko & Spiller, 2014; SGS, 2014). For luxury goods in general, the results confirm the existence of a perceived individual, functional, social and financial facet (Wiedmann, Hennnigs, & Siebels, 2009). To our knowledge, the perceived definition of luxury food has not been empirically investigated.

The studies are introduced by an empirical investigation with 220 test persons that reveals **expectations of consumers towards upscale food products**. The results are compared to influencing factors on the market for common food products. An explorative principal component reduces 16 items to four factors. The aim was to provide an empirical first approach to

---

[34] For theoretical information on hierarchical partial least squares path models see Becker, Klein, and Wetzels (2012).

consumer behavior in the market for luxury foods. The latter is here defined by the luxury attribute *price*. This decision is based on the advantage of *price* that its level is objectively to ascertain, in contrast to other factors that influence the perceived luxury value of a good (Wiedmann, Hennigs, & Siebels, 2007, 2009; Dubois, Laurent, & Czellar, 2001). The existence of the postulated trends on food markets toward more indulgence, quality and sustainability are addressed by this introductory study. Furthermore, it is aimed at providing impulses for sensibly designing the research design of the studies on the perceived dimensions of luxury food. The analysis shows that consumers expect a prestige value, trustability and sustainability as well as indulgence and a better taste when they buy upscale food products. For common food products, consumers are influenced in their purchase decisions by marketing, trustability and sustainability as well as the price-performance ratio. Interestingly, the factor price does not matter for upscale food products. We further found that consumers give high priority to quality and indulgence in their nutrition and are widely willing to pay higher prices for higher levels of both. The results are thus congruent with the findings of e.g. Kirig and Ruetzler (2007), Nestlé Deutschland AG (2012) and Padilla Bravo et al. (2013), who show that food quality and sustainability aspects do increasingly matter on German food markets. Especially the factor on trustability and sustainability, that includes four items on upscale food products, displays the relevance of *new luxury dimensions* in high-price segments on German food markets and inspire the following representative studies on the perceive dimensions of luxury food.

The results of the **study on segmenting German consumers based on perceived dimensions of luxury food** imply that the values of luxury food correspond to the values of luxury in general. What we found to be different is the intersection of the individual dimension with two others. In one factor, hedonism is combined with the social dimension in terms of prestige. In another one, self-identity is conjoined with a preference for special food qualities. Overall, the individual dimension, as conceptualized by Wiedmann, Hennnigs, and Siebels (2009), is found to define luxury food value in terms of three specifications, the third is materialism. The social aspect of luxury food is concentrated on prestige, while functionality matters in terms of usability, quality and uniqueness. The financial dimension in terms of price also applies in the perceptions of consumers towards food, but it is not as valid. The explained share of overall variance is the smallest among all factors and the Cronbach's Alpha value for the factor Price is the weakest in the principal component analysis.

In addition to those dimensions, it could be shown here that a further dimension of *new luxury values* in terms of sustainability and authenticity values is significant in the context of luxury foods. This result is consistent with both, literature on foods and literature on luxury. Honkanen, Verplanken, and Olsen (2006), SGS (2014) and Padilla Bravo et al. (2013) find that aspects of sustainability and authenticity are recently concerned by food consumers in industrialized countries and correspondingly, the demand for organic and Fair Trade food products as well as animal welfare is increasing. For luxury markets, Beverland (2005, 2006), Bilharz and Belz (2008), Joy et al., (2012), Kapferer (2010) as well as Meurer and Manninger (2012) discuss the increasing relevance of sustainability and authenticity as new luxury values. According to the latter authors, authenticity, regionalism and so called Green Luxury even belong to the most important trends in the international market for luxury brands.

In the cluster analysis, consumer segments are built upon using the revealed dimensions of luxury food. It shows that we can sensibly differentiate among four consumer segments. Two of them are recognized as target groups for the market of premium foods, which is associated with high quality and high prices. One group shows the highest mean value for the dimension of hedonism and prestige while in the other, uniqueness is most perceived as a dimension of luxury food. In both groups, the willingness to pay for luxury food is high compared with the other segments. Another segment contains the target group for sustainable and authentic food products. This group is least willing to pay for luxury food and finds luxury food rather unimportant. This is the biggest among all four segments, it contains 29.9% of the participants. Finally, one cluster is highly materialistically oriented but not be expected to buy premium food brands due to its comparatively low income. Nevertheless, it may have a high affinity toward marketing strategies that gear themselves toward functional luxury values or self-identity for food products of low or ordinary price ranges.

In the **partial least squares path model analysis** of the second study on luxury foods, we found that all dimensions from the first study significantly contribute to a perceived luxury food value. Self-identity/Quality has the highest path coefficient (0.187), closely followed by Usability, Uniqueness (both 0.186) and Materialism (0,185), Prestige/Hedonism and Sustainability/Authenticity are next ranging between (0.170; 0.168), and Price has the weakest estimated coefficient (0.017). Again, the financial dimension turns out to be less valid for the perceived definition of luxury food. The comparison of squared correlations for example show that price is the only dimension that correlates only very weakly with the latent construct. Among all tested luxury dimensions, functionality in terms of usability, uniqueness and quality contributes most to a luxury food value from a consumer´s point of view. This finding is completely congruent with new trends in German food consumption that Lueth et al. (2004), Kirig & Ruetzler (2007), Nestlé Deutschland AG (2012) observed. Accordingly, the change is from less price sensitivity toward an emphasis on high quality, specialties consumption and healthy meals.

Overall, the studies show that luxury food is perceived to be a multi-faceted and subjective construct. Wiedmann, Hennigs, and Siebels (2007, 2009) show the same thing for luxury in general. The motives for the consumption of luxury food are manifold and their particular weightings differ among segments in the market. We further confirm that modern consumption motives, in this case sustainability and authenticity, are significant determinants of perceived definitions of luxury food.

The second study revealed that the other dimensions, except price, have a stronger causal effect on the perceived luxury value than *new luxury values*. However, this could be due to the fact that luxury and sustainability is still perceived as opposites in some consumer segments (Kapferer, 2010; Padilla Bravo et al., 2013). Interlinking luxury with sustainability concerns is a new phenomenon in industrialized countries (Halaszovich & Meurer, 2012; Pruene, 2013). It is possibly a too early stage of development for providing stronger evidence for the association of a luxury value with sustainability. Monitoring the significance of sustainability as dimension of luxury food in future time periods would thus be a valuable next research step.

Table 1 gives an overview over the revealed luxury consumption motives and luxury definitions in horse sports and the market for luxury food or upscale food products.

Table 1. Key results on consumption motives in the markets for horse sports and luxury food

Study	Object of research	Target group and luxury market under investigation	Motives for luxury consumption or luxury dimensions
Luxusaffinität deutscher Reitsportler – Implikationen für das Marketing im Reitsport	The affinity of German equestrian people toward a material understanding of luxury	German equestrian people (non-representative sample); Luxury goods in general and horse sports	• Immaterial, hedonic and functional motives and the desire for luxury experience are predominant • A purely material definition of luxury as well as the motive of prestige are only relevant in a small segment • Doing horse sports is often seen as a luxury

Combining One-To-One Marketing and High-End Luxury: Theory-building from Customized Luxury Saddles for Chinese horse riders	The attitude of Chinese equestrian people toward customized luxury riding saddles	Chinese equestrian people (representativeness of the sample cannot be tested); Luxury goods in general and riding saddles	• Uniqueness, quality and superior craftmanship • Hedonic motives and self-expression • Prestige and conspicuousness are less relevant • Doing horse sports is mostly associated with intangible values
Introductory study: Luxusmarketing bei Lebensmitteln: Eine empirische Studie zu Dimensionen des Luxuskonsums in Deutschland	Expectations of German consumers toward upscale food products and factors that influence purchase descisions on the market for common foods	German consumers (non-representative sample); upscale food products	• Prestige • Vertrauen und Nachhaltigkeit • Sensorik und Hedonismus
Segmentation of German Consumers based on Perceived Dimensions of Luxury Food	The perceived dimensions of luxury food and consumer segmentation	German consumers (representative sample); Luxury food	• Prestige and hedonism, • Self-identity and quality, • Sustainability and authenticity, • Materialism, • Usability, • Uniqueness and • Price are perceived dimensions of luxury food
The Significance of definitional Dimensions of Luxury Food	The significance of the dimensions of luxury food that are revealed before	German consumers (representative sample); Luxury food	• All perceived dimensions of luxury food (see above) are significant • Functional and individual luxury dimensions have the strongest effects on a luxury food value

In summary, the first four chapters of the dissertation contribute to research by empirically testing the application of traditional and modern consumption motives in three luxury markets. Thereby, we investigated tangible luxury goods (riding saddles and luxury or upscale foods) as well as *luxury experience* (horse-riding activities).[35] We find evidence for an overall shift of motives for luxury consumption and luxury definitions away from prestige and conspicuousness. But the results also imply that motives for luxury consumption differ and depend on the particular luxury market. The studies show that marketers are advised to differentiate between *luxury experience* and tangible luxury goods. The first is strongly associated with immaterial values like having fun and fulfilling oneself, while for saddles and luxury food, quality, usability and uniqueness are repeatedly identified as significant consumption motives. They appear to be superficial for consumers of tangible luxury goods. In the case of luxury food, the results further show that the social component of luxury, in terms of prestige, still belongs to its perceived dimensions. Van der Veen (2003), who emphasizes the social meaning of luxury food consumption, finds similar results.

Marketing strategies should therefore be adapted to the motives of particular luxury markets, whereas the definition of the latter should base on the differentiation between tangible luxury and *luxury experience*. This seems to be more effective than emphasizing cultural specificities, since the influence of cultural backgrounds on motives for luxury consumption appears to be not as significant in our research as postulated in earlier research (e.g. Wong & Ahuvia, 1998). The study of Chinese horse riders shows that this target group is oriented toward Western values and has similar attitudinal and socio-demographic characteristics as German horse riders. The convergence of motives for luxury consumption among cultures could be an outcome of increasingly globalized markets, where luxury appears to be a prime example. Chinese luxury consumers, for example, are often active travelers, 63% of them have bought at least a few luxury goods outside of mainland China in 2012 (Atsmon et al, 2012).

A cross-cultural approach is also followed by Wiedmann, Hennigs, and Siebels (2007, 2009), who draw similar conclusions from their results: They state "Even though consumers in different parts of the world buy or wish to buy luxury products for apparently varied reasons, they possess similar values and—regardless of their country of origin—their basic motivational drivers are really the same: the financial, functional, personal, and social dimensions of luxury value perceptions, only the individual weighting differs" (Wiedmann, Hennigs, & Siebels, 2007, p. 9).

In order to give accordant practical recommendations, the contributions of Atwal and Williams (2008) as well as Tsai (2005) should be mentioned. They discuss the shift of motives for luxury consumption toward personal experience. Luxury customers are taken as emotional

---

[35] As inspired by Atwal and Williams (2009) and Meurer and Hirschsteiner (2012).

actors in the market who want to participate in the creation of luxury values. Marketing strategies are advised to offer sensory and emotional luxury experiences and inter-activities to their target groups. According to Pine and Gilmore (1998), this can be done by customer entertainment or providing aesthetic, escapist or educational experience. However, what Atwal and Williams (2008) suggest is the application of experiential marketing strategies to material luxury goods. The emotional, personally-oriented nature of *luxury experience* should be added to general luxury brands. Marketing is thus faced with the challenge to conjoin tangible luxury with *luxury experience*. Following the results of our studies, this implies that immaterial values are combined with functional motives (uniqueness, quality and functionality) that we found in the markets for riding saddles and luxury food.

Figure 1 visualizes the concluding remarks of the last two paragraphs. Thai's (2005) work, which explains the luxury motive of independent self-construal and its influence on luxury brand repurchase intention, inspires this model. The model of Thai (2005) is expanded in accordance with our results. It refers to the subdivision into motives for the consumption of *luxury experience* and motives for the consumption of tangible goods, and shows the intersection between both. Furthermore, it refers to the still existent but significance loosing socially-oriented luxury motives, prestige and conspicuousness. Our studies revealed that these motives rather appear in the consumption of tangible luxury goods than in the consumption of *luxury experience*. Even though, our results do not provide evidence that prestige and conspicuousness are not present among consumers of *luxury experience*. Socially-oriented luxury consumption is thus illustrated as being detached from the categorization.

For further research, the model's validity should be empirically tested. Therefore, we recommend to first prove the existence of converging luxury consumption patterns among various cultures, second, to validate the differentiating characteristics between the markets for tangible versus intangible luxury goods and third, to empirically prove the efficiency of marketing strategies that combine motives for *luxury experience* with tangible luxury goods.

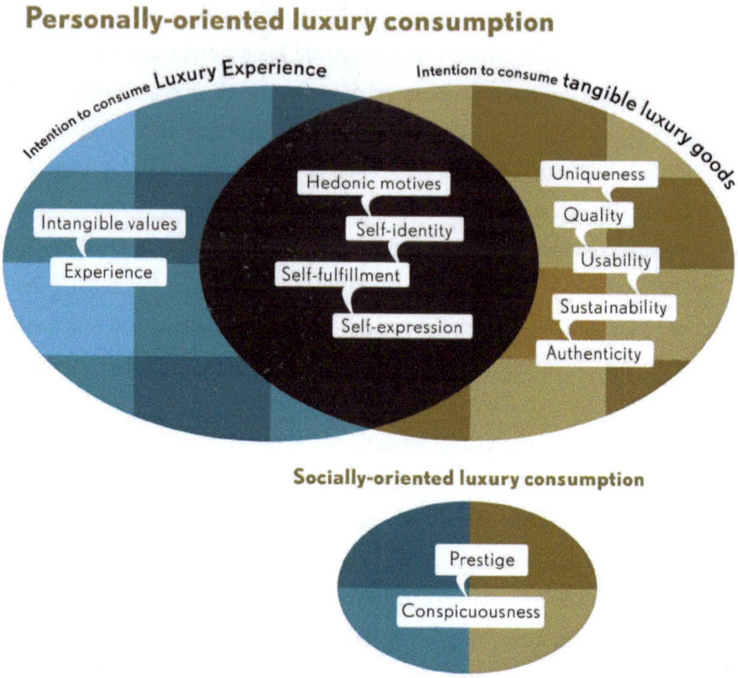

Figure 1. Motives for luxury consumption (own diagram)

The **excursion** contains two studies on horse sports markets. The aim was to investigate recent marketing issues in a luxury-associated branch. The **first study** critically analyzes the current ranking systems of the German and of the international association for equestrian athletes (*Fédération Nationale, FN,* and *Fédération Equestre Internationale, FEI*). Based on the results, improvements are proposed and tested for both rankings of the fifty most successful jumping horses in the world by means of an experiment. The ranking of sports horses is often taken as a basis for various decisions in equestrian sports and the market for horses, e.g. the purchase or sale of horses, the choice for studs and stallions in horse breeding and the nomination of rider-horse combinations for teams, special competitions and championships. Thus, it is used as one of the few objective valuation methods for horses and rider-horse combinations. According to Mamerow (2010), Arnemann (2003) and Arnold (2010), the performances of horses in horse sports is dependent on multiple parameters. For example, keeping, feeding, the horses' character, the horses' states of training, the weather, the competitors and the harmony between the horse and the rider influence the results on a competition. Against this background, a ranking system in horse sports should incorporate as many of the influential variables as possible while the content of subjectivity in the estimation of those variables has to be kept small. On the other hand, riders, breeders and investors in the horse market expect it to be transparent and easily applicable. If these requirements are not fulfilled, the original aim of a ranking in sports, which is to compare competitors and to increase the transparency of performances (Stoelting, 2002), cannot be reached. Testing the validity of the existing ranking system and its constant improvement should therefore be a crucial concern.

The rankings of the FN and FEI are cumulative, but the ranking of the FEI is more differentiated in the allocation of points (FEI, 2013; FN, 2013). We show that the shift from a purely cumulative ranking toward a more differentiated system improves its validity. We propose a ranking system that is based on average points and takes into account plus or minus points for good but not awarded performances or early withdrawals in a competition. We find that this system better predicts the results of the CHIO (Aachen, Germany), which is one of the most relevant international competitions for horse sports worldwide, than a cumulative ranking. Furthermore, we discuss that a cumulative system favors horses that compete more often compared to others. The focus is on the quantity of tournaments entries, while an averaging system puts emphasize on the quality of every single successful entry. The latter system tends to estimate higher scores for horses that get higher places on fewer events than horses that get lower places on more events. It creates an incentive for riders to put emphasize on a particular performance and not on the number of starts. Therefore, the appliance of the proposed averaging ranking is expected to improve the quality and the horse-friendliness of the competitive horse sports.

In summary, we find that an averaging, differentiated ranking can better serve the demands of riders, breeders and investors in the horse market. The first study of the dissertation shows that the group of German horse riders and the target group for luxury markets intersect. German horse riders are segmental luxury affine and they have a high average income compared to the German population. Furthermore, some characteristics of horses as goods on equestrian markets remember on typical characteristics of luxury goods. Horses are traded under a high involvement of sellers and buyers. Purchases and sales are usually accompanied by emotions and finally, every particular horse is unique in the market.[36] Thus, if we assume that horse riders are driven by modern motives for luxury consumption, which are, among other motives, quality and sustainability, the new ranking system offers improvements in the eyes of consumers. By means of the discussed incentives set by averaging rankings, the quality of particular rides will be enhanced and riders will use their sport horses in a more sustainable way.[37] There is no reason to overstrain horses in order to take part on as many competitions as possible. Tournament entries with a bad result or abandoning will just lower the horses ranking. Overall, the proposed new ranking can be seen as a particular marketing method to address the value change on luxury markets.

The **second study** of the excursion empirically reveals what operators in the market for German equestrian tourism perceive as success factors of their business. An exploratory factor analysis is used to define the success factors, a linear regression analysis determine the causal effects of the validated factors on a perceived level of business success. The data base stems from a total of 107 operators who offer touristic riding activities to their customers. We built a subsample with 34 participants, who are specialized in riding activities for hiking customers. These operators are investigated separately because the combination of riding and hiking is an upcoming trend in equestrian tourism (Ikinger et al., 2013). Three perceived success factors can be found. These are control and labeling of high quality standards by official associations in horse sports, the satisfaction of operators with their own business and the image they perceive to have, and the perceived satisfaction of customers. In the overall sample, the perceived satisfaction of customers primarily explains the perceived level of success. In the subsample, control and a perceived positive image significantly affect the perceived level of success. Overall, our results implicate that customer orientation is a crucial determinant of success in the horse sports tourist sector. Its significance is also confirmed for businesses in general (Fornell, 1992; Kanji & Wallace, 2000). Customer satisfaction positively influences customer loyalty, which turns into a company's success. For the market of *luxury experience*, to which equestrian tourism can be assigned, evidence for those success factors comes from hospitality and restaurants (Oh, 1999; Wu & Liang, 2009). Our findings thus provide further support for

---

[36] See Atwal & Williams (2009), Wiedmann, Hennigs, and Siebels (2009) and Wong, Polonsky, and Garma (2008) for the discussion of high involvement and emotional purchases as well as uniqueness as typical elements in general luxury markets.

[37] Sustainability is one of the new luxury dimensions currently discussed in the literature (Beverland, 2005; 2006; Bilharz & Belz, 2008; Joy et al., 2012; Kapferer, 2010).

270

the marketing strategies in horse sports that focus on building relationships toward customers and enhancing their trust and loyalty.

References

Alléres, D. (1993). L'univers du luxe. *Regards sur l'actualité, 187*, 3-26.

Arnemann, S. (2003). *Haltung von Sportpferden unter besonderer Berücksichtigung der Leistung.* Hannover: Tierärztliche Hochschule Hannover.

Arnold, D. (2010). *Pferdewirtprüfung [Bd. 2]: Nachhaltige Fütterung.* BoD-Books on Demand.

Atsmon, Y., Dixit, V., & Wu, C. (2011). Tapping China's luxury-goods market. *McKinsey Quarterly*, 1-5.

Atsmon, Y., Ducarme, D., Magni, M., & Wu, C. (2012). The McKinsey Chinese Luxury Consumer Survey *Luxury Without Borders: China's New Class of Shoppers Take on the World.* McKinsey & Company.

Atwal, G., & Williams, A. (2009). Luxury brand marketing–the experience is everything!. *Journal of Brand Management, 16*(5), 338-346.

Becker, J. M., Klein, K., & Wetzels, M. (2012). Hierarchical latent variable models in PLS-SEM: guidelines for using reflective-formative type models. *Long Range Planning, 45*(5), 359-394.

Bellaiche, J., Kluz, M., Mei-Pochtler, A., & Wiederin, E. (2012). Luxe Redux: raising the bar for selling of luxuries. *Boston Consulting Group, Boston.*

Beverland, M. (2006). The 'real thing': branding authenticity in the luxury wine trade. *Journal of Business Research, 59*(2), 251-258.

Beverland, M. B. (2005). Crafting brand authenticity: the case of luxury wines. *Journal of Management Studies, 42*(5), 1003-1029.

Bilharz, M., & Belz, F. M. (2008). Öko als Luxus-Trend: Rosige Zeiten für die Vermarktung „grüner" Produkte?. *Marketing Review St. Gallen, 25*(4), 6-10.

Chevalier, M., & Lu, P. X. (2010). *Luxury China: Market opportunities and potential.* John Wiley & Sons.

Degen, R. J. (2009). Opportunity for luxury brands in China. *Journal of Brand Management, 6*(3/4), 75-85.

Dubois, B., & Laurent, G. (1994). Attitudes toward the concept of luxury: An exploratory analysis. *Asia-Pacific Advances in Consumer Research, 1*(2), 273-278.

Eisenhardt, K. M. (1989). Building theories from case study research. *Academy of management review,* 14(4), 532-550.

FEI (2013, 1. Feb.). Jumping Rules. 24th edition. URL: http://www.fei.org/sites/default/files/JumpRules_24thEd2013_mark-up_updated_1.2.13.pdf. Last accessed 10th April 2013.

FN (2013). *Leistungs-Prüfungs-Ordnung 2013. Regelwerk für den deutschen Turniersport.* Warendorf: FN-Verlag.

Fornell, C. (1992). A national customer satisfaction barometer: the Swedish experience. *Journal of Marketing, 56*(1), 6-21.

Halaszovich, T., & Meurer, J. (2012). Green Luxury–Chancen und Herausforderungen für eine nachhaltige Führung von Luxusmarken. In C. Burmann, V. König, & J. Meurer (Hrsg.), *Identitätsbasierte Luxusmarkenführung* (S. 155-165). Springer Fachmedien Wiesbaden.

Heinemann, G. (2008). *Motivations for Chinese and Indian consumers to buy luxury brands* (Doctoral dissertation, Auckland University of Technology).

Hofstede, G. (1993). Cultural constraints in management theories. *The Academy of Management Executive, 7*(1), 81-94.

Honkanen, P., Verplanken, B., & Olsen, S. O. (2006). Ethical values and motives driving organic food choice. *Journal of Consumer Behaviour, 5*(5), 420-430.

Ikinger, C., Münch, C., Wiegand, K., & Spiller, A. (2013). Reiterleben, Reiterwelten: Zielgruppen zwischen Reitweisen, Motiven und der Liebe zum Pferd. Georg-August-Universität Göttingen, HorseFuturePanel UG, Dietz und Consorten (Eds.). URL: http://www.uni-goettingen.de/de/document/download /1988e74b5e6a7bf92bf38381a71a47f0.pdf/2013-04%20reitsportstudie_screen.pdf. Last accessed: 20 January 2015.

Institut für Demoskopie Allensbach (2013): AWA Allensbacher Markt- und Werbeträgeranalyse 2013, Allensbach.

Joy, A., Sherry, J. F., Venkatesh, A., Wang, J., & Chan, R. (2012). Fast fashion, sustainability, and the ethical appeal of luxury brands. *Fashion Theory: The Journal of Dress, Body & Culture, 16*(3), 273-296.

Kanji, G. K., & Wallace, W. (2000). Business excellence through customer satisfaction. *Total Quality Management, 11*(7), 979-998.

Kapferer, J. N. (2010). All that glitters is not green: The challenge of sustainable luxury. *European Business Review*, 40-45.

Kirig, A., & Rützler, M. H. (2007). Food-Styles. *Die wichtigsten Thesen, Trends und Typologien für die Genuss-Märkte*. Zukunftsinstitut-Studie. Kelkheim.

Lee, D. Y., & Dawes, P. L. (2005). Guanxi, trust, and long-term orientation in Chinese business markets. *Journal of International Marketing*, *13*(2), 28-56.

Lu, P. X. (2008). *Elite China: luxury consumer behavior in China*. John Wiley & Sons.

Lueth, M., Spiller, A., Enneking, U. (2004): *Analyse des Kaufverhaltens von Selten- und Gelegenheitskäufern und ihrer Bestimmungsgründe für/gegen den Kauf von Öko-Produkten*. Projektabschlussbericht für das BMVEL im Rahmen des Bundesprogramms ökologischer Landbau, Goettingen.

Mamerow, A. (2010). *Das Pferd ist dein Spiegel: Besser reiten durch mentales Training*. Leipzig: Draksal Verlag.

Meurer, J. (2012). Ebony or Ivory–wie glänzend ist die Zukunft des Luxus in Deutschland? Kritische Reflexionen zum Luxusmarkenmanagement. In C. Burmann, V. König, & J. Meurer (Hrsg.), *Identitätsbasierte Luxusmarkenführung* (S. 321-336). Springer Fachmedien Wiesbaden.

Meurer, J., & Hirschsteiner, S. (2012). The Art of Luxury Experience–Customer Experience Management zur erfolgreichen Umsetzung von Luxusmarkenerlebnissen. In C. Burmann, V. König, & J. Meurer (Hrsg.), *Identitätsbasierte Luxusmarkenführung* (S. 201-220). Springer Fachmedien Wiesbaden.

Meurer, J., & Manninger, K. (2012). Quo vadis globale Luxusmarkenführung–Status, Trends und Top-Themen für die CMO-Agenda. In C. Burmann, V. König, & J. Meurer (Hrsg.), *Identitätsbasierte Luxusmarkenführung* (S. 13-31). Springer Fachmedien Wiesbaden.

Nestlé Deutschland AG (2012) (Hrsg.). *Nestlé Studie 2012, Das is(s)t Qualität*, Auszüge aus der Nestlé Studie 2012.

Nitzko, S., & Spiller, A. (2014). Zielgruppenansätze in der Lebensmittelvermarktung. In M. Halfmann (Hrsg.), *Zielgruppen im Konsumentenmarketing* (S. 315-332). Springer Fachmedien Wiesbaden.

Oh, H. (1999). Service quality, customer satisfaction, and customer value: A holistic perspective. *International Journal of Hospitality Management*, *18*(1), 67-82.

Padilla Bravo, C., Cordts, A., Schulze, B., & Spiller, A. (2013). Assessing determinants of organic food consumption using data from the German National Nutrition Survey II. *Food Quality and Preference*, *28*(1), 60-70.

Peppers, D., & Rogers, M. (1993). *The One to One Future: Building Relationships One Customer at a Time*, New York: Currency/Doubleday.

Peppers, D., & Rogers, M. (1997). *The one to one future*. New York: Doubleday.

Peppers, D., Rogers, M., & Dorf, B. (1999). Is your company ready for One-to-one marketing. *Harvard Business Review*, 77(1), 151-160.

Pine, B. J., & Gilmore, J. H. (1998). Welcome to the experience economy. *Harvard Business Review*, 76, 97-105.

Pruene, G. (2013). *Luxus und Nachhaltigkeit*. Springer Fachmedien Wiesbaden.

Schuette, H., & Ciarlante, D. (1998). *Consumer behavior in Asia*. New York: New York University Press.

SGS (2014). Vertrauen und Skepsis: Was leitet die Deutschen beim Lebensmitteleinkauf? SGS-Verbraucherstudie 2014: Ergebnisse einer bevölkerungsrepräsentativen Befragung. Hamburg: SGS Germany GmbH.

Silverstein, M. J., Fiske, N., & Butman, J. (2008). *Trading Up: why consumers want new luxury goods--and how companies create them*. Penguin.

Stoelting, E. (2002). Wissenschaft als Sport: Ein soziologischer Blick auf widerspuechliche Mechanismen des Wissenschaftsbetriebes. *die hochschule, 2*, 58-78.

Tsai, S. (2005). Impact of personal orientation on luxury-brand purchase value. *International Journal of Market Research, 47*(4), 427–452.

Van der Veen, M. (2003). When is food a luxury?. *World Archaeology, 34*(3), 405-427.

Wiedmann, K. P., Hennigs, N., & Siebels, A. (2007). Measuring consumers' luxury value perception: a cross-cultural framework. *Academy of Marketing Science Review, 7*(7), 333-361.

Wiedmann, K. P., Hennigs, N., & Siebels, A. (2009). Value□based segmentation of luxury consumption behavior. *Psychology & Marketing, 26*(7), 625-651.

Wong, C. Y., Polonsky, M. J., & Garma, R. (2008). The impact of consumer ethnocentrism and country of origin sub-components for high involvement products on young Chinese consumers' product assessments. *Asia Pacific Journal of Marketing and Logistics, 20*(4), 455-478.

Wong, N. Y., & Ahuvia, A. C. (1998). Personal taste and family face: Luxury consumption in Confucian and Western societies. *Psychology and Marketing, 15*(5), 423-441.

Wu, C. H. J., & Liang, R. D. (2009). Effect of experiential value on customer satisfaction with service encounters in luxury-hotel restaurants. *International Journal of Hospitality Management, 28*(4), 586-593.

Yeoman, I. (2011). The changing behaviours of luxury consumption. *Journal of Revenue & Pricing Management, 10*(1), 47-50.

Zhan, L., & He, Y. (2012). Understanding luxury consumption in China: consumer perceptions of best-known brands. *Journal of Business Research, 65*(10), 1452-1460.

Limitations and Outlook

Three overall limitations of the dissertation should be discussed. With regard to a systematic investigation of modern motives for luxury consumption, the dissertation is limited to sustainability, authenticity, hedonic motives and self-identity in both studies on luxury food. Indeed, questions about the general attitudes towards luxury, open questions and the findings from interviews and observations on the food market and in horse sports let us conclude for other new luxury motives like immaterial values and indulgence. Nevertheless, the need for next research steps on the further development and appliance of approaches to systematically analyzing the value change in luxury consumption are desirable. In the studies on luxury food, an instrument is proposed that turns out to be sensibly employed for the investigations of motives. Wiedmann, Hennigs, and Siebels (2007, 2009) inspire the instrument. We extend it to specificities of the market for luxury food. The actuality of the change in luxury consumption and the pressing need for research is stated in the marketing literature (Meurer & Manniger, 2012).

Furthermore, the representativeness of the studies on equestrian athletes as well as of the study on upscale food products is restricted. In the German survey of equestrian atheletes, our sample is at least biased towards higher levels of education, a higher share of people living in a rural area and lower income levels compared to the entire population of German riders (Institut für Demoskopie Allensbach, 2013). Compared to the entire German population, we primarily observe a sample bias toward higher levels of education and younger ages (Statistisches Bundesamt, 2013). For the Chinese survey on equestrian athletes, the comparability with an entire sample of Chinese horse riders is not possible at present due to the lack in availability of an appropriate data set. However, what can be taken for granted are levels of education and income in this target group which are far above the Chinese averages (National Bureau of Statistics of China, 2012). We also find a high level of involvement and enthusiasm of participants in their horse sports. We can assume that at least the levels of income and education as well as the share of people doing serious, competitive sports are higher in our sample than in the entire group of Chinese horse riders. Finally, the sample for the German survey on upscale food products is biased toward younger ages, higher levels of education and lower levels of income compared to the entire German population (Institut für Demoskopie Allensbach, 2013; Statistisches Bundesamt, 2013). Employing representative samples for further investigations on the motives of horse riders and consumers in the market for upscale food products in Germany and internationally would test the validity of our results.

Finally, the dissertation is focused on only two luxury-associated agricultural markets. Its results should thus not be confused with universally clarifying the phenomenon of modern luxury consumption. In literature, the existence of various markets where new luxury motives may significantly influence the behavior of consumers is postulated (Bellaiche et al., 2012; Hagtvedt & Patrick, 2009). Empirical investigations are hardly found on consumption motives of particular new emerging luxury markets. Against this background, this dissertation (only) offers a reference point for further empirical studies on consumption motives in today's luxury markets. It is desirable that the value change of luxury consumption is systematically investigated in various luxury markets. Thereby, studies should focus on the categorization of luxury consumption in tangible luxury goods and *luxury experience*.

References

Bellaiche, J., Kluz, M., Mei-Pochtler, A., & Wiederin, E. (2012). Luxe Redux: raising the bar for selling of luxuries. *Boston Consulting Group, Boston*.

Hagtvedt, H., & Patrick, V. M. (2009). The broad embrace of luxury: Hedonic potential as a driver of brand extendibility. *Journal of Consumer Psychology, 19*(4), 608-618.

Institut für Demoskopie Allensbach (2013): AWA Allensbacher Markt- und Werbeträgeranalyse 2013, Allensbach.

Meurer, J., & Manninger, K. (2012). Quo vadis globale Luxusmarkenführung–Status, Trends und Top-Themen für die CMO-Agenda. In C. Burmann, V. König, & J. Meurer (Hrsg.), *Identitätsbasierte Luxusmarkenführung* (S. 13-31). Springer Fachmedien Wiesbaden.

National Bureau of Statistics of China (2013). China Statistical Yearbook 2012. URL: http://ebook.dgbas.gov.tw/public/Data/3117141132EDNZ45LR.pdf. Last accessed 02. February 2015.

Statistisches Bundesamt (2013). Gesellschaft und Staat. URL: https://www.destatis.de/DE/ZahlenFakten/GesellschaftStaat/StaatGesellschaft.html. Zugriff am 27. Juli. 2013.

Wiedmann, K. P., Hennigs, N., & Siebels, A. (2007). Measuring consumers' luxury value perception: a cross-cultural framework. *Academy of Marketing Science Review, 7*(7), 333-361.

Wiedmann, K. P., Hennigs, N., & Siebels, A. (2009). Value□based segmentation of luxury consumption behavior. *Psychology & Marketing, 26*(7), 625-651.

Curriculum Vitae

Laura Helena Hartmann

Date of Birth	October 20, 1986
Nationality	German

Education

01/2013 – 05/2015	**Doctoral Studies in Agricultural Economics**

Georg-August-University of Goettingen (Germany),

Institute of Agricultural Economics and Rural Development,

Marketing for Foods and Agricultural Products

Mentoring Professor: Prof. Dr. Achim Spiller

Title of the Dissertation: *Consumption Motives in Luxury Marketing*

– An Analysis of two Agricultural Markets:

Equestrian Sports and Food –

09/2009 – 09/2011	**Master's Program in Economics and Economics**

University of Bern (Switzerland)

Completed with *Master of Science in Business and Economics*

Main focus on Economics

Master's Thesis in the area of Microeconomics with the title:

The Peculiar Economics of Equestrian Sports

10/2006 – 08/2009	**Bachelor's Program in Economics**

Carl-von-Ossietzky-University of Oldenburg (Germany)

Completed with Bachelor of Arts in Economics

Main focus on Economics

Bachelor's Thesis in the area of Political Economy with the title:
Unilateral Climate Policies and Carbon Leakage

08/1999 – 06/2006	**German university entrance qualification with the advanced courses of German and Politics**

Ubbo-Emmius-Gymnasium in 26789 Leer / Germany

Practical Experience

01/2013 – 05/2015	**Activities as a scientific assitant at the Georg-August-University of Goettingen**

- Organization of a conference on Marketing of Luxury Goods with Human Consumption at the Georg-August-University of Goettingen in May 2015
- Mentoring of Bachelor's and Master's Theses as well as the preparation and support of the seminar execution on *Sports Marketing* (Prof. Dr. Spiller) in winter semester 2013/2014
- Execution of a market study for a manufacturer of luxury goods in summer 2013

01/2012 – 08/2012	**Internship with Allianz Managed Operations and Services SE in Munich**

Member of a Human Resource – Project Team within the IT-division of the Allianz-subsidiary

Activities:

- Producing drafts to optimize the selection process of *High Potentials*
- Preparation and execution of job interviews

- Representation of the company on company fairs as well as the organization of events for students
- Maintenance of existing and development of new co-operations with national and international universities
- Reconditioning and design of communication concepts
- Writing a scholarly article on Technology Acceptance Models

| 08/2008 – 09/2008 | **Internship with United Bulk Carriers in Wayne/USA** |
| | Versatile tasks in the field of *Operative Shipping Company Business* |

| 03/2008 – 03/2008 | **Internship with the Deutsche Bank AG in Singapore** |
| | Operative tasks in the field of *Private Wealth Management* |

Languages

German	Mother tongue
English	Very good
French	Basics
Latin	Acquisition of the *Latin Certificate*

Computer Literacy

Microsoft Office	Very good
SPSS	Very good
SmartPLS	Good
EViews	Good
Matlab	Basic
Stata	Basic

Extramural Engagement

05/2006 – dato   **Semi-professional Dressage riding**

- Work as instructor of dressage horses
- Successful participation in national & international tournaments
- Holding classes of riding lessons (class C – training license for competitive sports with distinction)
- Breeding of young horses as well as engagement as Hanoverian horses breeder
- Documentation of judges' marks on big equestrian tournaments
- Engagement in students' riding groups in Oldenburg and Bern during local enrollment periods

03/2007 – 10/2008   **Member of Youth Dressage Squad of Lower Saxony**

Paper and Presentations

Refereed journals

Hartmann, L. H., Schulz-Wiemann, C., Spiller, A., & Gertheiss, J. (2014). Weiterentwicklung der Rankingsysteme im Reitsport (in abgeänderter Version). *Sportwissenschaft, 44*(2), 99-115.

Sidali, K. L., Eggemann, M., Hartmann, L., Filaretova, O., & Dörr, A. C. (2013). Success Factors of Equestrian Tourism: Evidence from Germany (in abgeänderter Version). Turistica Gen - Giu 2013, 73-83.

Hartmann, L., Kerssenfischer, F., Fritsch, T., & Nguyen, T. (2013). User Acceptance of Customer Self-Service Portals. *Journal of Economics, Business and Management, 1*(2), 150-155.

Conference proceedings

Hartmann, L. H., & Spiller, A (2014). Combining One-to-One marketing and High-End Luxury: Theory-building from Customized Luxury Saddles for Chinese Horse Riders (Short version). *Proceedings of the 2014 Global Marketing Conference at Singapore. Bridging Asia and the World: Globalization of Marketing & Management Theory and Practice.* ISSN number: 1976-8699.

Published discussion paper

Hartmann, L. H., & Spiller, A. (2015). Luxusaffinität deutscher Reitsportler - Implikationen für das Marketing im Reitsportsegment (in abgeänderter Version). Diskussionspapier 1501. *Diskussionspapiere der Georg-August-Universität Göttingen, Department für Agrarökonomie und Rurale Entwicklung,* ISSN 1865-2697.

Schneider, T., Hartmann, L. H. & Spiller A. (2015). Luxusmarketing bei Lebensmitteln: Eine empirische Studie zu Dimensionen des Luxuskonsums in der Bundesrepublik Deutschland (in abgeänderter Version). Diskussionspapier 1502. *Diskussionspapiere der Georg-August-Universität Göttingen, Department für Agrarökonomie und Rurale Entwicklung,* ISSN 1865-2697.

Paper for submission

Hartmann, L., Nitzko, S., & Spiller, A. (2014). Segmentation of German Consumers based on Perceived Dimensions of Luxury Food.

Hartmann, L., & Spiller, A. (2015). The Significance of Definitional Dimensions of Luxury Food.

Praxis-oriented work

Hartmann, L, & Jaeschke, M. (2013): The Equestrian Market Place in China: Market Entry Strategy for Hermés Equestrian Products. Conducted in collaboration with *Equestrian Globe GmbH* (Germany), on behalf of *Hermés* (France).

Eckjans, S., & Hartmann, L. (2013). Neue Märkte im Pferdesektor–China. *Goettinger Pferdetage ´13: Zucht, Haltung und Ernährung von Sportpferden. Tagungsband Goettinger Pferdetage 2013.* 33-35.

Presentations

2014 Global Marketing Conference at Singapore (July 2014). Presentation of the paper *Combining One-to-One Marketing and High-End Luxury: Theory-building from Customized Luxury Saddles for Chinese Horse Riders.*

AG Pferd (January 2014). Presentation *Horse Sports in China - Results and Impressions of a Research Journey to Beijing and Shanghai in Summer 2013.*

Lecture "Sportmarketing" (held by Prof. Dr. Achim Spiller, Georg-August-University of Goettingen, in winter semester 2013/2014) (December 2013). Presentation of the paper *Weiterentwicklung der Rankingsysteme im Reitsport – ein Experiment* (Sportwissenschaft, 44(2), 99-115).

Acknowledgements

Am Ende möchte ich die Gelegenheit nutzen, den Menschen einen großen Dank auszusprechen, die mir während meiner Promotionszeit zur Seite gestanden und mich unterstützt haben.

Ich blicke auf zwei schöne und spannende Jahre zurück, die für mich ein Leben lang die Bedeutung eines sehr wichtigen und prägenden Lebensabschnitts einnehmen werden.

Zunächst danke ich meinem Doktorvater Prof. Dr. Achim Spiller für die gute Betreuung meiner Dissertation. Obwohl ich durch meine externe Promotion keinem Projekt angeschlossen war, hat er für alle meine Studien großes Interesse aufgebracht, mich motiviert und inspiriert und mich stets in die Abläufe am Lehrstuhl integriert. Ich habe mich dort als festes Mitglied des Lehrstuhl-Teams verstanden gefühlt und das außerordentlich gute, vertrauensvolle Arbeitsklima sehr genossen. Durch seine Art, uns Doktoranden großes Vertrauen entgegen zu bringen und selbstständiges, eigenverantwortliches Arbeiten zu fördern, gibt er uns die Möglichkeit, sich in der eigenen Persönlichkeit zu entfalten und von Interessen leiten zu lassen.

Einen besonderen Dank möchte ich ihm auch dafür aussprechen, dass er mir die Möglichkeit gegeben hat, eine Tagung zu meinem Promotionsthema zu organisieren.

Weiterhin danke ich Herrn Prof. Dr. Klaus-Peter Wiedmann für die Übernahme des Zweitgutachtens sowie für das inspirierende Gespräch zur Modellbildung im Luxusgütermarketing.

Herrn Prof. Dr. Ludwig Theuvsen danke ich für das Übernehmen der Funktion des Drittprüfers meiner Arbeit.

Ein Dank geht ebenso an meine Kollegen und Freunde am Lehrstuhl, mit denen ich viel schöne Zeit in und auch außerhalb der Uni verbracht habe. An dieser Stelle möchte ich auch Petra Geile dafür danken, dass sie uns Doktoranden tagtäglich nicht nur in unserer Arbeit unterstützt, sondern auch immer ein offenes Ohr für jegliche Freuden, Probleme und Sorgen hat.

Herrn Sacha Eckjans und Frau Maiken Jäschke danke ich dafür, dass sie mir ermöglichten, eine Studie im chinesischen Reitsport durchzuführen.

Meiner Familie danke ich von Herzen für ihre vielfältige Unterstützung während meiner gesamten Studien- und Promotionszeit. Als Konstante in meinem Leben erweisen sie mir einen so wertvollen Dienst, für den es mir fast unmöglich erscheint, angemessen Danke zu sagen. Durch meine Familie wurden mir Studium und Promotion und mein Engagement im Reitsport, das mir den nötigen Ausgleich verschaffte, überhaupt erst ermöglicht.

All meinen Freunden danke ich dafür, dass sie für mich da sind und mir durch viele schöne, unterhaltsame Treffen und gemeinsame Unternehmungen sowie gute Gespräche den Alltag versüßen.

Ein herzlicher Dank gilt schlussendlich meinem lieben Jan für die seelische Unterstützung und seine wertvollen fachlichen Ratschläge.